NATO ASI Series

Advanced Science Institutes Series

A series presenting the results of activities sponsored by the NATO Science Committee, which aims at the dissemination of advanced scientific and technological knowledge, with a view to strengthening links between scientific communities.

The Series is published by an international board of publishers in conjunction with the NATO Scientific Affairs Division

A	Life Sciences	Plenum Publishing Corporation
B	Physics	London and New York
C	Mathematical and Physical Sciences	Kluwer Academic Publishers
D	Behavioural and Social Sciences	Dordrecht, Boston and London
E	Applied Sciences	
F	Computer and Systems Sciences	Springer-Verlag
G	Ecological Sciences	Berlin Heidelberg New York
H	Cell Biology	London Paris Tokyo Hong Kong
I	Global Environmental Change	Barcelona Budapest

PARTNERSHIP SUB-SERIES

1. Disarmament Technologies	Kluwer Academic Publishers
2. Environment	Springer-Verlag
3. High Technology	Kluwer Academic Publishers
4. Science and Technology Policy	Kluwer Academic Publishers
5. Computer Networking	Kluwer Academic Publishers

The Partnership Sub-Series incorporates activities undertaken in collaboration with NATO's Cooperation Partners, the countries of the CIS and Central and Eastern Europe, in Priority Areas of concern to those countries.

NATO-PCO DATABASE

The electronic index to the NATO ASI Series provides full bibliographical references (with keywords and/or abstracts) to about 50000 contributions from international scientists published in all sections of the NATO ASI Series. Access to the NATO-PCO DATABASE compiled by the NATO Publication Coordination Office is possible in two ways:

- via online FILE 128 (NATO-PCO DATABASE) hosted by ESRIN,
  Via Galileo Galilei, I-00044 Frascati, Italy.

- via CD-ROM "NATO Science & Technology Disk" with user-friendly retrieval software in English, French and German (© WTV GmbH and DATAWARE Technologies Inc. 1992).

The CD-ROM can be ordered through any member of the Board of Publishers or through NATO-PCO, Overijse, Belgium.

2. Environment – Vol. 1

The Partnership Sub-Series incorporates activities undertaken in collaboration with NATO's Cooperation Partners, the countries of the CIS and Central and Eastern Europe, in Priority Areas of concern to those countries.

The volumes published as a result of these activities are:

Vol. 1: Clean-up of Former Soviet Military Installations. Edited by R. C. Herndon, P. I. Richter, J. E. Moerlins, J. M. Kuperberg, and I. L. Biczó. 1995

Vol. 2: Cleaner Technologies and Cleaner Products for Sustainable Development. Edited by H. M. Freeman, Z. Puskas, and R. Olbina. 1995

Clean-up of Former Soviet Military Installations

Identification and Selection of Environmental Technologies for Use in Central and Eastern Europe

Edited by

Roy C. Herndon
John E. Moerlins
J. Michael Kuperberg

Institute for Central and Eastern
European Cooperative Environmental Research
Florida State University
2035 East Paul Dirac Drive, Suite 226 Morgan Building
Tallahassee, Florida 32310-3700, USA

Peter I. Richter
Imre L. Biczó

Center for Hungarian/American Environmental Research
Studies and Exchanges
Technical University of Budapest
Budafoki ut 8.
H-1111 Budapest, Hungary

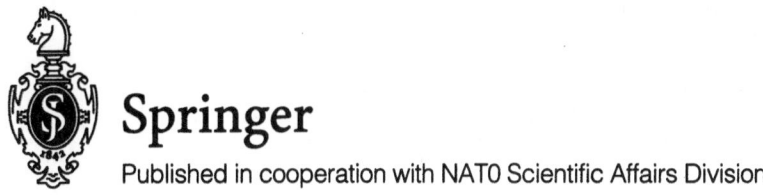

Published in cooperation with NATO Scientific Affairs Division

Proceedings of the NATO Advanced Research Workshop "Identification and Selection of Technologies for Use at Former Soviet Military Installations in Central and Eastern Europe", conducted in Visegrad, Hungary, June 21–23, 1994

ISBN 978-3-642-63361-4   ISBN 978-3-642-57803-8 (eBook)
DOI 10.1007/978-3-642-57803-8

CIP data applied for

© Springer-Verlag Berlin Heidelberg 1995
Originally published by Springer–Verlag Berlin Heidelberg New York in 1995
Softcover reprint of the hardcover 1st edition 1995

Typesetting: Camera-ready by authors
SPIN: 10475207   31/3136 – 5 4 3 2 1 0 – Printed on acid-free paper

PREFACE

A NATO Advanced Research Workshop (ARW) was conducted on June 21-23, 1994 in Visegrád, Hungary related to the clean-up of former Soviet military installation sites. This ARW included a technical site visit to the Komárom Base Site which is a former Soviet military installation in Hungary. During this three-day ARW, a strategy and set of recommendations were developed for selecting technologies and evaluating remediation approaches for these sites. This strategy incorporated such critical issues as the economic and financial conditions of the region, temporal considerations with regard to the urgency for which remedial actions are needed for these sites, the prioritization of resource allocations for site clean-up using risk-based considerations, and other crucial issues which will affect the implementation of remedial activities in the region. Approximately 40 invited experts, representing a number of different disciplines as well as both NATO and Cooperation Partner countries from the region, participated in this ARW.

The types of former Soviet military installations in Central and Eastern Europe include: aircraft bases, fueling areas, maintenance and repair facilities, training grounds, non-ammunition storage areas (for lubricants, chemicals, paints, equipment), ammunition storage areas, medical facilities, production facilities, and municipal facilities. Environmental contamination at these sites poses significant human health and environmental risks. Site contaminants include: solvents (e.g., BTEX), mineral oil hydrocarbons, polycyclic aromatic hydrocarbons (PAHs), chlorinated hydrocarbons, heavy metals, pesticides residues, and polychlorinated biphenyls (PCBs). The primary environmental media adversely affected by these contaminants are soils, ground water and surface water.

The principal objective of this ARW was to evaluate approaches for remediating the contamination problems at these sites in Central and Eastern Europe by convening a group of international experts on site remediation and related disciplines. The ARW results also have direct application to other sites in the region with similar contamination problems. A key recommendation made by the ARW participants relates to the need to establish a technology demonstration site in the region. The purpose of this demonstration site would be to provide first-hand knowledge and experience in the actual application and effectiveness of these remediation technologies. The demonstration project could also provide training for both agency personnel and private sector personnel involved with contaminated site remediation activities in the region.

The editors wish to acknowledge the guidance provided by Dr. Luis Veiga da Cunha concerning the organization, format and focus of the workshop. In addition, the contributions of the NATO Advisory Panel on the Priority Area of the Environment were greatly appreciated during the workshop by providing an effective context for the ARW participants to conduct their work. We would also like to express our appreciation to the workshop participants whose expertise, dedication and hard work made the workshop a success. The editors would like to recognize the efforts of the following individuals for their diligence and care in the preparation of this document: Reyn C. Anderson, Laymon L. Gray, Loreen K. Kollar, and Joseph R. Shaeffer.

Table of Contents

Country Summaries

RECOMMENDATIONS FOR REMEDIATION ACTIVITIES AT

FORMER SOVIET MILITARY INSTALLATIONS IN CENTRAL AND

EASTERN EUROPE

Roy C. Herndon, John E. Moerlins and J. Michael Kuperberg
Center for Hungarian/American Environmental Research, Studies and Exchanges
Florida State University
2035 East Paul Dirac Drive, Suite 226 Morgan Building
Tallahassee, Florida 32310-3700
USA

Peter I. Richter and Imre L. Biczó
Center for Hungarian/American Environmental Research, Studies and Exchanges
Technical University of Budapest
Budafoki ut 8.
H-1111 Budapest
Hungary

Introduction

The principal objective of this NATO Advanced Research Workshop (ARW) was to evaluate approaches for remediating the contamination problems at former Soviet military installations in Central and Eastern Europe by convening a group of international experts on site remediation and related disciplines. Many of these sites are located in the Czech Republic, Hungary, Poland and the Slovak Republic. Representatives from each of these countries as well as various environmental restoration experts from both NATO countries and non-NATO countries (including the Cooperation Partner Countries identified by NATO) participated at this workshop. These experts included individuals who have specialties in site characterization, environmental monitoring, public health evaluation, as well as treatment and related site remediation technologies. Also included were individuals from government agencies, research institutes and universities, as well as private sector environmental engineering and consulting companies. The mix of experts from the various countries, disciplines and areas of expertise provided the basis for productive discussions of the pressing issues associated with site remediation needs at these former Soviet military sites.

NATO ASI Series, Partnership Sub-Series, 2. Environment – Vol. 1
Clean-up of Former Soviet Military Installations
Edited by R. C. Herndon et al.
© Springer-Verlag Berlin Heidelberg 1995

The overall mission of the workshop was to develop a set of recommendations, and a rationale for implementing these recommendations within the context of the short-term and long-term needs of the countries within the region. A primary conclusion reached by the ARW participants was that the two fundamental problems facing the region are the limited financial resources and, to some extent, a lack of experience in utilizing some of the more innovative and, in some cases, less costly remediation approaches and technologies. In addition, there is generally a lack of experience in using risk assessment and related evaluations for ranking sites, prioritizing resource allocations, and for making decisions related to selecting remediation approaches for these sites. In order to address these concerns and implement the ARW findings, the following recommendations were developed:

- It is recommended that the initial site assessment/characterization phase should include a comprehensive audit of conditions at the site, identification of areas of potential concern with regard to further characterization activities and an accumulation of information regarding site uses and activities.

- It is recommended that site characterization technologies that have not been extensively utilized in Central and Eastern Europe should be investigated and demonstrated at appropriate sites (e.g., at the Komárom Base, Hungary). While some of the more advanced site characterization technologies have been applied at sites in Central & Eastern Europe, experience differs among countries and no mechanism exists to exchange information on these technologies. Some recognized and effective site characterization technologies have not been broadly applied in Central & Eastern Europe (e.g., photoacoustic infrared and Fourier-transform infrared spectrometry).

- It is recommended that remediation technologies that have not been extensively utilized in Central and Eastern Europe should be investigated and demonstrated at appropriate sites (e.g., at the Komárom Base, Hungary). Many of the identified remediation technologies have been applied at one or a few Central & Eastern European sites. While some of the more advanced remediation technologies have been applied at sites in Central & Eastern Europe, experience differs among countries and no mechanism exists to exchange information on these technologies. Some recognized and effective remediation technologies have not been broadly applied in Central & Eastern Europe (e.g., soil washing, *in-situ* bioremediation).

- It is recommended that a database be developed containing the history and experiences of the various former Soviet military installations in the Central and Eastern European countries. Such a database would allow future investigators to make use of information gathered from past experiences in other countries. The problems associated with contamination are likely to be similar from site to site but will be specific to activities that were conducted at a given site (e.g., airbase, maintenance depot).

- It is recommended that, with regard to cleanup levels, governmental responsibility primarily should be to remediate to levels that avoid imminent risks to human health. Other considerations concerning initial remedial actions should include the migration of contaminants off-site, adverse effects to non-human receptors and the cost associated with taking no action.

- It is recommended that target cleanup levels for guidance of remedial actions should be based on the proposed land use, and should be developed according to established toxicology and risk assessment procedures.

- It is recommended that training be provided to potential users in the region on toxicology and risk assessment procedures, in order to more appropriately and more cost-effectively prioritize sites and evaluate remediation approaches for cleaning up former military installations and other sites with contamination in the region.

- It is recommended that future land use (and associated risks) should be matched with the nature and extent of contamination and the remediation approach selected for a site. A generic matrix of possible site contamination types and proposed uses should be created to guide the evaluation of proposed remediation approaches for a site.

- It is recommended that a process for selecting an appropriate remediation strategy for a site should proceed using the following sequence of activities:
  Problem definition
  Establishment of objectives
  Development of alternatives
  Analysis and comparison of alternatives
  Implementation of selected alternative
  Monitoring of progress and completion of selected alternatives

- It is recommended that a process for the analysis of alternatives and selection of specific technologies for site remediation should proceed according to the following criteria (order does not indicate relative importance of criteria):
  Short-term effectiveness
  Long-term effectiveness
  Long-term reliability
  Implementability
  Cost

- It is recommended that, to the extent possible, selection of a preferred site remediation approach should proceed using a standardized methodology which quantifies the selection criteria in a weighted fashion. It may be useful to retain a flexible component in the context of allowing evaluation of qualitative factors.

- It is recommended that funds from the sale or transfer (privatization) of former Soviet military installations should be earmarked for cleanup of other (possibly less marketable) sites.

- It is recommended that an EnviroTRADE Network Implementation (Phased) Plan be developed. This would include a permanent regional node at the Technical University of Budapest and primary nodes at strategic locations in the Central European region. This would also include secondary nodes at selected site locations.

- It is recommended that an EnviroTRADE Network System be provided for the proposed nodes. Support to be provided would include hardware, installation, training, data entry, system maintenance and upgrades.

- It is recommended that a specific demonstration project at a representative site (e.g., at the Komárom Base, Hungary) be implemented. The location and timing of this project should be such that all interested individuals from Central and Eastern Europe have the opportunity to participate. Participation should involve both the demonstration of selected technologies, including Central and Eastern European technologies and training in the application of these technologies.

The final recommendation, to establish a centrally located and easily accessible demonstration site in the region, forms the foundation for addressing many of the more fundamental issues related to the remediation of former Soviet military installations. It was determined, based on the discussions at the ARW, that there is generally not a deficiency of innovative technologies available to the region. Innovative site remediation technologies are available worldwide and documentation describing the implementation, use and comparisons of these, and the more traditional technologies (e.g., pump and treat, or thermal treatment technologies), is also widely available throughout the region. However, what is lacking in the region with regard to remediation technologies (including all aspects of site remediation: characterization, monitoring, treatment, risk evaluation, etc.) is first-hand knowledge and experience with the implementation and use of these technologies and related evaluations. The purpose of the recommended demonstration site is to provide this first-hand knowledge and experience to decision-makers in the region and to train officials at all levels of government on the use and evaluation of these remediation technologies.

The focus of these recommendations is not to address a deficiency in technical expertise but rather to address a need to provide experience, information and training that can be used to help decision-makers in the region to prioritize sites and evaluate remediation approaches for cleaning up former Soviet military installations. The enhancement of knowledge for indigenous decision-makers would result in more efficient (i.e., faster, less costly and, better) site remediation choices regarding activities which are generally expensive, and which hold great significance with regard to human health, the quality of the environment, and the overall long-term economic prosperity of the region.

This workshop also provided a secondary benefit in that the knowledge and experience obtained from the ARW can be readily transferred to address clean-up problems at other sites (i.e., those not associated with former Soviet military installations). It also could be readily expanded to address other environmental problems in the region in facilitating more cost-effective environmental management procedures.

The information contained in this book provides a comprehensive picture of the site remediation needs and the approaches being implemented within Central and Eastern Europe at former Soviet military sites. The workshop clearly demonstrated that significant opportunities for international cooperation exist both for countries within the region and among countries from within and from outside of the region. It is through activities such as this NATO ARW that the seeds of these opportunities for international cooperation can bear fruit.

STRATEGY FOR IDENTIFYING AND EVALUATING SITE

REMEDIATION APPROACHES FOR FORMER SOVIET MILITARY

BASES IN CENTRAL AND EASTERN EUROPE

Charles F. Voss
Golder Federal Services, Inc.
4044-148th Avenue NE
Redmond, Washington
USA

1. INTRODUCTION

The Central and Eastern European (CEE) countries that hosted the former Soviet military installations during the past 48 years are now facing the difficult situation of resolving the environmental legacy associated with the operation of these bases. There are thousands of contaminated sites at the former Soviet military installations in Hungary, Poland, the Czech Republic, and the Slovak Republic. There are 171 former Soviet military installations in Hungary alone with at least 20 designated as high-priority sites. The sites include artillery ranges, training grounds, aboveground and underground storage areas and tanks, fueling areas and maintenance and repair areas. They are contaminated with hydrocarbons, heavy metals, acids, PCBs, PAHs, solvents, ordnance, biological agents and numerous other wastes that have contaminated the soil, surface water, and groundwater.

Little or no financial responsibility for the remediation of these bases has been accepted by the former tenants and the countries in which the bases are located are now facing the issue of how to respond to the potential hazards present at the bases. This situation, coupled with the challenges of transition from the former centrally planned economic system to the new political, social, and economic reality requires that the environmental remediation efforts be focused on the sites that represent the most direct risks to human health and the environment. The need for assistance in these efforts is recognized by the Regional Environmental Center for Central and Eastern Europe, which stated that "The methodological and managerial capacity of environmental decision makers must be improved through extensive training and

NATO ASI Series, Partnership Sub-Series, 2. Environment – Vol. 1
Clean-up of Former Soviet Military Installations
Edited by R. C. Herndon et al.
© Springer-Verlag Berlin Heidelberg 1995

advising. A comprehensive step-by-step procedure for approaching environmental problems needs to be formulated and put forward."

Most importantly, the first step in addressing the environmental problems associated with the former Soviet military bases is to assess the risk to human health and the environment represented by each of the bases. Clearly, some of the military bases pose a greater hazard because of the types and levels of contaminants present and their proximity to populated areas. Unfortunately, the level of remedial investigation that has been carried out is very preliminary in many cases and the associated health risks are poorly defined. The financial resources for such activities are rather limited in part because economic problems such as unemployment and decreases in real wages are common in Central and Eastern Europe and are being given priority over environmental issues.

Once a decision has been made to remediate a site, the selection of a remedy is often difficult because of complex site conditions, the large number of remedial alternatives that exist, and, in some instances, the lack of experience at cleaning up some types of contaminants. Under these situations, it is useful to employ a decision-making process for selecting a remedy for a site. This paper attempts to provide insight into how to go about selecting a remediation approach for a given site. The methodology involves a series of activities or steps that have evolved over the last decade to select remedial alternatives for contaminated sites in the United States (U.S.).

2. PROPOSED STRATEGY

The five components of the strategy are fundamental to problem solving. The steps are to: (1) define the problem; (2) establish objectives; (3) develop alternatives; (4) compare alternatives; and (5) implement the solution. This general process for selecting a remedy is used routinely in the U.S. and is based largely on the methodology developed by the U.S. Environmental Protection Agency (EPA) for remedial investigations under the Comprehensive Environmental Response, Compensation, and Liability Act (CERCLA, also known as Superfund) (EPA, 1988). Each of the five components are discussed below and an example is provided to illustrate the process.

2.1 Define the Problem

The first step in the methodology is to define the problem in terms of the sources of contamination and to predict how the contaminants can be transported (i.e., release mechanisms and pathways) to human beings and the environment (i.e., receptors). The level of effort required will depend on the complexity of the site (i.e., the number and types of contaminants present, the areal extent of contamination, and the number and types of pathways). This conceptual model becomes the basis for evaluating the potential risk to human health and the environment as well as for developing and evaluating alternative remediation approaches. It can also be used to prioritize remediation activities when a decision maker is faced with multiple sites. This important use of risk assessment is addressed in other papers presented during this workshop.

Typical sources of contamination are drums, storage tanks, waste piles, landfills and surface impoundments containing hazardous substances. Heavily contaminated soils are often considered as sources, especially when the original source (e.g., leaking pipeline) no longer exists. Pathways for transporting contaminants from the source to the potential exposure points include, for example, groundwater migration, airborne transport, and biological uptake. Information that is normally collected includes surface features (e.g., buildings, tanks, piping), geology, soils, surface water hydrology, hydrogeology, meteorology, human populations, proposed land use, and ecology.

Some level of site characterization (problem definition) has been completed at many of the former Soviet military installations in Central and Eastern Europe although the amount and quality of information varies considerably between sites and countries. The types of activities included review of aerial photographs, visual inspection and cataloging of known and potential sources (e.g., fuel tanks, waste pits), geophysical surveys, and soil and groundwater analysis. Historical information about the operation of many of the installations has been difficult to obtain, especially for bases that were occupied exclusively by Soviet military personnel. Some of the early attempts to assess the ecological damages were impeded by the local Soviet authorities.

Information on the source, location and physical site data have been used to provide a preliminary estimate of the location of contaminants. Groundwater monitoring programs have been initiated at many of the most seriously contaminated sites. The primary sources of contamination include:

- petroleum derivatives containing various hydrocarbons and derivatives floating on the surface of the groundwater;

- residual fuels, solvents, etc., in underground and aboveground storage tanks as well as in the underlying soil;

- unlined disposal pits containing a wide variety of hazardous waste; and

- chemical warfare agents in old storage sites or field test sites.

The principal pathways for transporting the contaminants are the groundwater and surface waters around the sites, with some agricultural lands also containing surface contamination. Residential areas that depend on the local aquifers and surface waters for drinking water or domestic use are located near many of the contaminated sites. The surface water bodies and drinking water wells often constitute the most likely routes for human exposure through ingestion of the water or food sources that have been contaminated by the water.

2.2 Establishing Remediation Objectives

Before alternatives for remediating a site can be compared and analyzed, it is first necessary to establish objectives for the remedial action. The objectives are aimed at protecting human health and the environment and are typically specified in terms of the contaminants concerned, the exposure routes (pathways) and receptors, and acceptable levels of exposure. Quantitative objectives are preferable since they provide a clear method for evaluating the effectiveness of a remedy.

Typical methods for establishing objectives include review of regulatory requirements (e.g., the maximum concentration of a specific contaminant in groundwater) and site-specific risk assessment. In some instances, federal or state regulations, standards, or requirements may exist that are applicable to the release or threatened release of a contaminant at a site. For example, the state administration authorities in the Slovak Republic established pollutant concentration limits in soils and groundwater that require restoration (Fatulová and Geisbacher, 1994). The concentration of non-polar extractable substances (NEL) was used as the measure for when a contaminated soil had to be remediated (NPES > 600 mg/kg) and the Czechoslovak drinking water standard was used for surface and groundwaters.

In the U.S., acceptable levels of carcinogenic and non-carcinogenic risk have been established for many substances, and these can be used as the preliminary remediation objectives. A risk assessment can be performed to determine whether the risk posed by a site for all source-pathway-receptor combinations exceeds the risk-based objective. If the target level is exceeded, remediation is required to reduce the risk to the target level. A remedy could meet such an objective by reducing the hazard (e.g., removing the source of contamination) or by limiting the exposure (providing an alternative drinking water source).

A remediation objective should not specify a technology but rather should focus on the contaminants of concern, exposure routes, receptors, or a measurable standard of performance or control. The protection of human health can be achieved by reducing exposure (e.g., limiting access to a site), as well as by reducing the contaminant levels. Examples of objectives are: (1) reduce the groundwater constituent levels to comply with drinking water standards; (2) increase the travel time for contaminated groundwater to 100 years (assumes natural degradation of the contaminant with time); and (3) carcinogenic risk $< 10^{-6}$. Ideally, the remedial action objectives should address both a target contaminant level and an exposure route, and not focus on contaminant levels alone.

2.3 Develop Alternatives

Once the remediation objectives have been established, alternatives for remediating the site can be developed using the following steps:

Develop General Response Actions - In this step, classes of remedial technologies are identified that will satisfy the remedial action objectives. Examples of classes of remedial technologies are treatment, containment, excavation, extraction, disposal, institutional controls, or some combination of these. The general response actions should be stated in broad terms (e.g., extract and treat contaminated groundwater) and should not necessarily specify how the water would be extracted or what treatment technology would be used.

Quantify the Volumes or Areas Involved - The areas or volumes of media associated with a general response action needs to be estimated. The objective of this step is to quantify the areal, and volumetric extent of the various media that will be remediated. For example, if the general response action is to remove volatile organic compounds from low permeability soil, it would be necessary to estimate the volume of the contaminated soil.

Identify and Screen Remedial Technologies - After the media associated with the general response actions have been characterized, technology types and process options can be identified and evaluated. Process options refer to categories of a remedial technology. For example, cutoff walls, recovery wells, and recovery trenches are process options for the response action "containment." The technologies are then screened based on their effectiveness, implementability and cost. This evaluation results in their retention or elimination from further consideration on the basis of one or more of these attributes. The technologies can be quickly screened by assigning a qualitative score (e.g., favorable, neutral, unfavorable) to each and then eliminating the less promising ones.

Assemble Technologies Into Alternatives - The final step is to combine the general response actions using different technology types and different volumes of media and/or areas of the sites. More than one general response action may be applied to a medium where the type and distribution of contaminants vary spatially and thus require a combination of processes (e.g., soil washing and capping). The goal is to develop a wide range of alternatives that can be compared in subsequent trade-off studies using additional criteria.

In refining the alternatives, it is important to realize that while protection is achieved by reducing exposures to acceptable target levels, it is not always possible to clean-up a specific medium to these levels and additional measures may be required to limit future exposures (e.g., it may not be possible to clean-up drinking water to acceptable levels and the protection may have to be provided by a wellhead treatment system).

EXAMPLE

An example that illustrates the above steps may be useful. Consider a hypothetical site containing a localized region of contaminated soil where some portion of the contaminants have leached and are present in the local groundwater. The maximum allowable groundwater concentration for the contaminant is assumed to be 10 ppm, which is used as the objective for the remedial action. Possible general response actions for the contaminated groundwater are listed below with the area or volume involved shown in parentheses:

- no action
- institutional controls (all residents in the affected area)
- collection/treatment (all groundwater with a concentration > 10 ppm)
- containment (areas where the groundwater has a concentration > 10 ppm)

General response actions for the contaminated soil include:

- no action
- institutional controls (all residents in the affected area)
- excavation/treatment (10,000 m^3)
- *in-situ* treatment (2,000 m^3)
- containment (1,000 m^2)

Table 1 contains a list of these general response actions and the associated technologies and potential process options. Table 1 also shows the result of the screening process and the rationale for eliminating several of the technologies. The shaded entries pertain to process options and technologies that were eliminated on the basis of effectiveness and/or implementability. The remaining technologies were combined to form a range of site-wide alternatives (Table 2). The alternatives include: (1) no action; (2) limited action; (3) removal of the source without groundwater treatment; (4) source containment without groundwater treatment; (5 and 6) source removal with groundwater collection and treatment; and (7 and 8) source containment with groundwater collection and treatment. The number of alternatives to be carried through the final stage of evaluation should be large enough to include the range of treatment and containment technologies initially presented, but typically does not exceed 10 for practical reasons.

Table 1. Example of Initial Screening of Technologies and Process Options

Groundwater General Response Actions	Remedial Technology	Process Options	Screening Comments
No Action	None	Not Applicable	Potentially applicable
Institutional Controls	Alternate water supply	City	Potentially applicable
		New community well	Potentially applicable
	Monitoring	Groundwater monitoring	Potentially applicable
Collection/Treatment	Extraction wells	Biological treatment	Not effective for the contaminant present
		Chemical treatment	Potentially applicable
	Subsurface drains	Interceptor trench/biological treatment	Not effective for the contaminant present
		Interceptor trench/chemical treatment	Potentially applicable
Containment	Vertical barriers	Slurry wall	Could not implement. Depth to water table too great
		Grout curtain	Not effective. Grout not compatible with contaminant
	Cap	Multi-media	Potentially applicable
		Concrete	Potentially applicable
Soil General Response Actions	**Remedial Technology**	**Process Options**	**Screening Comments**
No Action	None	Not Applicable	Potentially applicable
Institutional Controls	Access restrictions	Fencing	Potentially applicable
Treatment	Excavation	Biological treatment	Not effective for the contaminant present
		Thermal treatment	Potentially applicable
	In Situ Treatment	Biological treatment	Not effective for the contaminant present
		Chemical fixation	Potentially applicable
Containment	Horizontal barrier	Grout injection	Difficult to implement because of the heterogeneity of the soil

2.4 Compare Alternatives

Once the range of alternatives is identified, they can be further analyzed to determine a preferred alternative. This is typically based on a comparative analysis using a set of criteria. The technology process options are evaluated with respect to the three screening criteria mentioned previously (effectiveness, implementability, cost) and one additional criterion, reliability, so that differences among alternatives can be identified. Reliability becomes an issue for the alternatives that leave untreated contaminants at the site (e.g., alternatives 2, 3, and 4 in Table 2).

The alternatives should be evaluated against both the short- and long-term aspects of their effectiveness, implementability and cost. Long-term is typically defined as 70 years or the average human life expectancy. The effectiveness of an alternative relates to its ability to provide overall protection of human health and environment and comply with applicable regulations or performance standards. Implementability refers to the feasibility of constructing, operating, and maintaining a remedial action from both a technical and managerial standpoint.

The selection of the preferred alternative can be made using an unstructured or structured evaluation process. In an unstructured evaluation, the alternatives are compared on a flexible basis using the following steps (LaGrega et al., 1994):
1. consider the interactions between the criteria (i.e., the conflicts and synergies);
2. identify the advantages and disadvantages of the alternatives relative to one another;
3. examine and balance the tradeoffs among the alternatives; and
4. reach a consensus on the preferred alternative.

The unstructured approach recognizes the subjective nature of assigning relative values to alternatives based on the criteria. Furthermore, the criteria are indeed relative considerations (i.e., they do not define a threshold below which an alternative is unacceptable). However, the success of building a consensus among groups responsible for a contaminated site, all with diverse interests and motivations, will rely heavily on the interpersonal skills of the person managing the evaluation. Differences of opinion concerning the advantages and disadvantages are possible and are often difficult to resolve.

In a structured process, the alternatives are ranked based on how they score on each of the evaluation criteria and the weight assigned to the different criteria (e.g., when cost is the overriding criterion it will have a higher weight). There are several methods for making such

Table 2. Combination of General Response Actions to Form Alternative Remedies

General Response Action		1 No Action	2 Limited Action	3 Source Removal; No GW Control	4 Source Containment No GW Control	5 Source Removal GW Collection, Treatment	6 Source Removal GW Collection, Treatment	7 Source Containment GW Collection, Treatment	8 Source Containment GW Collection, Treatment
Medium	Technology								
Ground water	Alternate water supply		◆	◆	◆				
	Monitoring	◆	◆	◆	◆	◆	◆	◆	◆
	Extraction/ Chemical Treatment					◆		◆	
	Trench/ Chemical Treatment						◆		◆
Soil	Access Restrictions		◆						
	Excavation/ Thermal Treatment			◆		◆	◆		
	Chemical Fixation				◆			◆	◆
	Cap				◆			◆	◆

a structured evaluation including the Analytical Hierarchy Process (Saaty, 1992) and Multiattribute Utility Theory (Von Winterfelt, 1986). A detailed discussion of such a process is outside the scope of this paper; however, the general approach is briefly illustrated below. Fundamentally, the process is very similar to and is compatible with the methodology for selecting a remedial approach described in this paper. The steps involved are:

1. define the problem - a clear statement of the decision problem must be formulated including the goal or objective as well as the factors that are important or can affect the decision

2. develop the decision structure - develop the hierarchy between the goal and the factors or criteria that are important for satisfying the goal

3. identify sub criteria under each criterion

4. judge the importance of each criterion - weights can be assigned to the criteria either in a quantitative way (if data on the criteria are available, e.g., cost) or subjectively by what is referred to as "pairwise comparison"

5. evaluate alternatives - each alternative is evaluated against the other based on the lowest criterion

6. check the reasonableness and sensitivity of the results - determine the robustness of the results (i.e., do small changes in the weighting factors change the decision?)

Figure 1 contains a layout of the decision structure. The goal is to protect human health and the environment by implementing one of the remedial alternative options in Table 2. The factors that are important for deciding the alternative to use are effectiveness, cost, implementability, and reliability. Two sub criteria, the short-term and long-term effectiveness of an alternative, are also considered in this example. A weight is assigned to each of the criteria based on their relative importance to the stated goal. For example, if cost is the principal consideration for selecting a remedy it would be assigned a relatively higher weight. Weights are also assigned to the sub criteria. The assignment of weights can be done by a panel or an individual depending on the situation. If there are differences in opinion on the relative importance of the criteria, several sets of weights can be assigned and analyzed to determine whether the differences in opinion change the outcome of the decision. This is a very effective means for addressing real or perceived differences among the various parties involved in the decision-making process.

Once weights have been assigned to the criteria, the alternatives are evaluated against one another based on each of the criteria to determine the preferred alternative. Inexpensive software packages that run on personal computers are available for performing such analyses and include sensitivity and other post-processing capabilities. Figure 2 shows the best five remedial alternatives from the example problem. (In this example, the weights assigned to the

Figure 1. Decision Structure for the Example Problem.

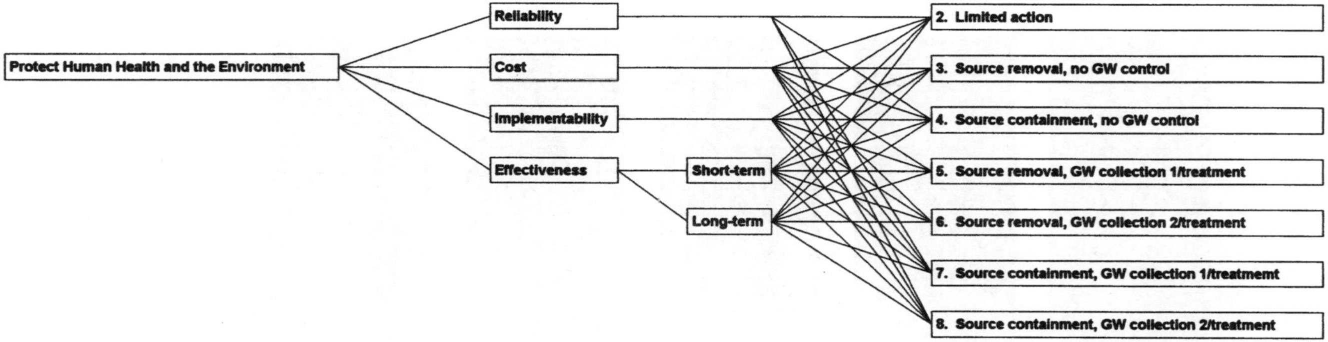

Figure 2. Ranking of the Top Five Remedial Alternatives from the Example Problem.

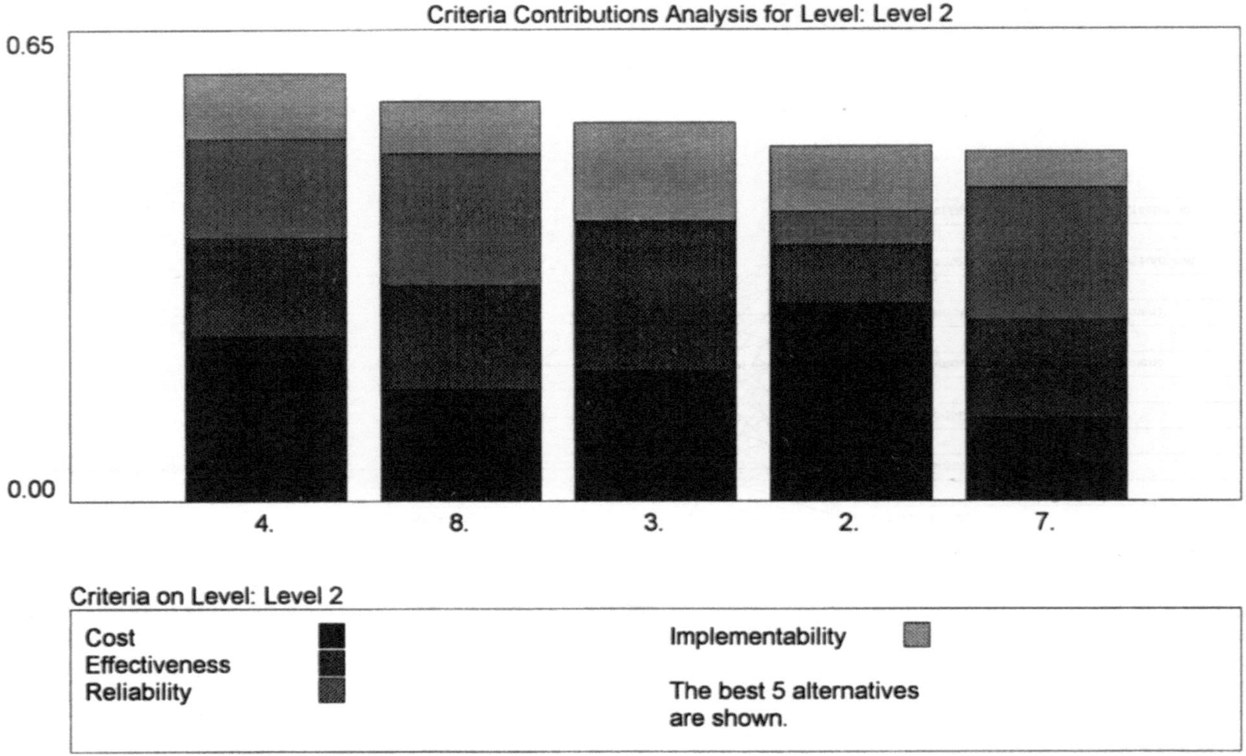

criteria and the relative scores for the alternatives are not provided. The scores are arbitrarily assigned for the purpose of illustrating the kind of output that typically is generated, in order to evaluate the reasonableness of the outcome and the robustness of the decision (recommended alternative)). The histogram was generated using one of the commercially available PC-based programs and illustrates the relative contribution of the various criteria in scoring the alternatives. Remedy 4 (source containment with no groundwater control) is the preferred alternative in this case based on the evaluation criteria. Utilizing the results of a sensitivity analysis, the preference is judged to be stable (i.e., the recommended alternative does not change with small changes to the criterion weights).

3. SUMMARY

The environmental challenges represented by the former Soviet military bases are formidable due to the large number of potentially hazardous sites and the limited resources for characterizing and remediating these sites. It is essential that efficient and effective procedures be adopted or developed to address these problems and provide adequate protection to human health and the environment over the short- and long-term. The selection of remediation approaches for these sites is only one aspect of the task at hand. A considerable amount of site characterization work is still needed in the region to provide a reliable basis for prioritizing the sites. Once this process has been completed, the methodology presented here can be used to identify and select remediation approaches.

4. ACKNOWLEDGMENTS

Preparation of this paper was supported in part by the International Technology Exchange Project at the U.S. Department of Energy Office of Technology Development under contract No. 87-3497 with Sandia National Laboratories. The technical and financial support provided by the U.S. Department of Energy and Sandia National Laboratories is greatly appreciated.

5. REFERENCES

Fatulová, E. and D. Geisbacher (1994), Environmental Aspects of Re-Using Former Soviet Army Bases in Slovakia, Presented at the NATO Advance Research Workshop in Viségrad, Hungary, June 21-23, 1994.

LaGrega, M.D., P.L. Buckingham, and J.C. Evans (1994), Hazardous Waste Management, McGraw-Hill, Inc., ISBN 0-07-019552-8.

Saaty, T.L. (1992), Decision Making for Leaders, Pittsburgh, RWS Publications.

The Regional Environmental Center for Central and Eastern Europe (1994), Strategic Environmental Issues in Central and Eastern Europe, Volume 1, Regional Report, ISBN 963 04 4304 X, REC, Budapest, Hungary, May, 1994.

U.S. Environmental Protection Agency (1988), Guidance for Conducting Remedial Investigations (1988), Feasibility Studies Under CERCLA, EPA/540/G-89/ 004, U.S. Department of Commerce, National Technical Information Service, Springfield, VA, October, 1988.

Von Winterfelt, D. and W. Edwards (1986), Decision Analysis and Behavioral Research, Cambridge University Press.

APPLICATION OF THE ENVIROTRADE INFORMATION SYSTEM

FOR THE CLEANUP OF A FORMER SOVIET MILITARY

INSTALLATION: THE KOMÁROM BASE SITE, HUNGARY°

Rudolph V. Matalucci, Mark W. Harrington, Charlene P. Harlan
Sandia National Laboratories
P.O. Box 5800, M/S 0743
Albuquerque, New Mexico 87185-0743
USA

J. Michael Kuperberg
Center for Hungarian/American Environmental Research, Studies and Exchanges
Florida State University
2035 East Paul Dirac Drive, Suite 226 Morgan Building
Tallahassee, Florida 32310-3700
USA

Imre L. Biczo
Center for Hungarian/American Environmental Research, Studies and Exchanges
Technical University of Budapest
Budafoki út 8
H-1111 Budapest
Hungary

Abstract

During a NATO Advanced Research Workshop (ARW) held in Visegrád, Hungary, June 21-23, 1994, portions of contamination data from the Former Soviet Union (FSU) site at Komárom, Hungary were used to demonstrate the international EnviroTRADE Information System as a tool to assist with the identification of alternative cleanup measures for contaminated sites. The NATO ARW was organized and conducted by the joint Florida State University and the Technical University of Budapest, Center for Hungarian-American Environmental Research, Studies, and Exchanges (CHAERSE). The purpose of the workshop was to develop a strategy for the identification and selection of appropriate low-cost and

---

°   This work was supported by the U.S. Department of Energy under contract No. DE-AC04-94-AL85000.
    SAND94-2275C.

NATO ASI Series, Partnership Sub-Series, 2. Environment – Vol. 1
Clean-up of Former Soviet Military Installations
Edited by R. C. Herndon et al.
© Springer-Verlag Berlin Heidelberg 1995

innovative site remediation technologies and approaches for a typical abandoned FSU site. The EnviroTRADE information system is a graphical and textual environmental management tool under development by the U.S. Department of Energy (USDOE) at Sandia National Laboratories (SNL) as a part of the cleanup program of the nuclear weapons complex. EnviroTRADE provides a powerful, multi-purpose, multi-user, multi-media, and interactive computer information system for worldwide environmental restoration and waste management (ER/WM). Graphical, photographic, and textual data from the Komárom site were entered into EnviroTRADE. These data were used to make comparative evaluations of site characterization and remediation technologies that might be used to clean up primarily hydrocarbon contamination in the groundwater and soil. Available hydrogeological and geological features, contaminated soil profiles, and topographical maps were included in the information profiles. Although EnviroTRADE is currently only partially populated (approximately 350 technologies for cleanup are included in the database), the utility of the information system to evaluate possible options for cleanup of the Komárom site has been demonstrated.

Introduction

Purpose: This paper highlights some of the applications of the EnviroTRADE Information System for potential site cleanup requirements and describes how the tool can be used to clean up a contaminated site near Komárom, Hungary. It describes the capabilities of EnviroTRADE as a tool for environmental scientists, engineers, and planners in developing a typical remediation effort.

Background: The U.S. Department of Energy (USDOE) is committed to remediation of waste sites throughout its complex by the year 2019. Examples of remediation problems are volatile organic compounds (VOCs) in soils and groundwater, soils contaminated with radionuclides and heavy metals, and mixed waste sites containing both hazardous and radioactive waste. Waste management problems include characterization, monitoring, treatment, and disposal of wastes. Environmental technologies must be identified and developed to facilitate the remediation of existing problems and minimize future waste streams.

DOE's Office of Technology Development (OTD) recognizes that it can accelerate the technology development efforts and enhance the expenditure of available funds through international cooperation among government entities, private industry, and educational institutions. Promising foreign technologies are being considered for application in the United States, and the USDOE hopes to involve the private sector in the application of U.S.

technologies abroad. Consequently, OTD is sponsoring the development of EnviroTRADE - an international information system that facilitates the exchange of environmental restoration (ER) and waste management (WM) problems and technologies worldwide [1], [2].

NATO Advanced Research Workshop: The purpose of the NATO Advanced Research Workshop is to prepare a technical report that recommends a procedure for cleaning up typical abandoned sites in the Former Soviet Union (FSU). This workshop is also supported by USDOE's Office of Environmental Restoration and Waste Management (EM), Office of Technology Development (OTD), under the International Technology Coordination Program (USDOE/EM-52.1), for cooperative activities on NATO site restoration and relevant technologies. The Center for Hungarian-American Environmental Research, Studies, and Exchanges (CHAERSE), established jointly by Florida State University and the Technical University of Budapest, are responsible for planning and organizing this NATO workshop.

The EnviroTRADE Information System is presented and described at this NATO workshop using data from an FSU military installation at Komárom, located about 40 kilometers from Budapest, Hungary. A demonstration EnviroTRADE workstation was installed by SNL at the Technical University of Budapest (TUB) during this NATO workshop. The system was populated with the Komárom Base site data, and the attending participants have been able to obtain first-hand experience operating the system for data retrieval and applying characterization and remediation technologies. The Komárom Base site will be visited by the workshop participants and used as a model for evaluating ER/WM problems and applying technological solutions. More than 40 technical experts representing numerous NATO (including the United States) and Central and Eastern European countries will make presentations to jointly develop generic procedures for ER/WM applications at some of the more environmentally critical abandoned FSU sites in the region.

EnviroTRADE Information System

System Description: The EnviroTRADE Information System contains profiles of domestic and foreign ER/WM problems and technologies [1]. It offers an alternative to searching manually through masses of written documents. The system's unique architecture and the easy-to-use interface allow the user to locate specific site descriptions or technology profiles by searches based on attributes such as waste type, soil conditions, and location of contaminants. Users are able to identify matches between worldwide problems and available or emerging technologies. For example, a user may browse through the information on "Available and Applied Technologies," select a particular technology for viewing, and then request that the system locate environmental problems that might have use for the technology. Conversely, the user may browse the information on "Environmental EM Problems," select a particular problem for viewing, and request that the system provide information and contacts for applicable technologies. Each technology profile includes performance information, cost and availability, technical contacts, and drawings and photographs. The system is able to compare candidate technologies from the point of view of effectiveness, availability, safety, public acceptance, and cost by the operator manually reading profiles and information. Where matches between problems and existing technologies cannot be found, the system identifies the potential for development of new and innovative technologies to address environmental problems. EnviroTRADE also provides general information on international energy and environmental organizations, sites, activities, and contacts.

EnviroTRADE is being developed on a Reduced Instruction Set Computing (RISC)-based UNIX workstation platform [1]. A relational database management system (RDBMS) is being used to store, manage, and retrieve information. A graphical user interface (GUI) communicates with the RDBMS to provide user-friendly "point and click" menus in a standard windowing environment. The system is easy to use and provides visually oriented information such as maps, photographs and diagrams of environmental sites and technologies, as well as text profiles of the problems and technologies. The prototype runs on a SUN workstation under the Open Look window manager as a stand-alone system. The system design is based on a client/server model that can be maintained and delivered over a network when the information base is large enough to require that architecture.

Capacity and Capability as an Environmental Tool: The EnviroTRADE Information System utilizes the capabilities of a geographical information system (GIS) viewer to manage and display any spatial data that might be associated with an entry in the database. When GIS views are available, a button appears in the GUI that allows the user access to the spatial information. This information might be in the form of a map that displays and describes environmental contamination at sites, in rivers, and over large land areas. Because the USDOE is funding the identification and restoration or management of sites within its complex, many of the national laboratories and other sites are capturing this spatial information using GIS software. These views, or appropriate subsets of views, can easily be embedded in the EnviroTRADE information base.

System Status and Potential Commercialization: The transfer of appropriate USDOE technologies that have the potential to benefit the private sector is a priority of the USDOE, and EnviroTRADE is one of the technologies being considered for transfer to industry. Commercialization involves identifying potential private investment partners, determining their level and nature of interest, and eliciting their guidance on the requirements that such a system must accommodate. If the current commercialization negotiations are successful, a private partner(s) would help develop and manage a commercial version of the EnviroTRADE system. Cooperative Research and Development Agreements (CRADAs) or other mechanisms would be used to formalize the relationships among the participants.

Future Plans and Programs: The USDOE recognizes the great need to have easy access to large volumes of accurate, current, affordable, and integrated information on environmental waste problems, technological solutions, and development efforts within the weapons complex. A long-term effort is needed to share and integrate environmental information and resources across the entire complex [2]. Information on waste sites must be organized and made accessible, and technologies within the USDOE and private sector matched to these problems. Regulations must be considered, and management tools made available.

An integrated system of data and decision support tools will also provide knowledge beyond matching of technologies to waste sites. It will allow for successes beyond the use of the raw data alone. Patterns in data may be discovered and intelligence can be applied to the information. Redundancy of cleanup efforts can be minimized and government funding used as effectively as possible. The USDOE/EM decision makers require methods for analyzing existing problems and technology development efforts to determine if their environmental needs are being addressed in the most efficient manner. Waste site managers, technology developers, and the private sector, including universities and other laboratories, need access to information about environmental research projects and performance of existing remediation technologies. In addition to these customers, the potential exists for use and expansion of the environmental information system by U.S. Department of Defense site managers and regulators such as the U. S. Environmental Protection Agency.

Komárom Site Database

Site Description and Former Use: In Central and Eastern Europe, hundreds of abandoned FSU military sites constitute a serious and widespread environmental problem. In Hungary alone, there are 170 of these abandoned sites. This environmental problem is aggravated by the economic difficulties of the countries in the region, allowing the application of only the most cost-effective remediation technologies. A proposed demonstration project would involve the comprehensive cleanup of one of these sites in order to demonstrate an effective and affordable approach to such activities [3]. The Hungarian Environmental Authority has recommended that the Komárom site be considered as a target site for a regional demonstration project.

The city of Komárom, near the northern border of Hungary, is the site of two abandoned former Soviet military bases, both of which show extensive contamination of soil and groundwater by various hydrocarbons associated with motor fuels. The Komárom site consists of two subareas: the Monostori Fort and the Arpad Army Post. Following the withdrawal of Soviet troops from Hungary, these military complexes were evaluated and found to be extensively contaminated and in poor states of repair [3].

The Monostori Fort was built during the last century by Italian army engineers. It is one of a number of military installations in a chain of forts built every 6-10 km along the Danube River to defend the Austro-Hungarian monarchy from possible attacks from the river. The Monostori Fort is one of the largest of the fortifications (58 hectares) associated with this defensive chain of military establishments. It is located approximately 150 m from the Danube River and is considered a valuable military-historical monument.

From the early 1950s until 1990, twelve (12) hectares of this facility were used by the FSU military as one of the largest ammunition storage areas in the region. Military activities associated with the use and subsequent abandonment of this site led to a large amount of hydrocarbon contamination. This has adversely affected groundwater resources and threatens the quality of the Danube River, which serves as the source of drinking water for cities along its course.

The Arpad Army Post was a garrison of approximately 28 hectares in the heart of the city of Komárom. The adjacent lands are densely populated residential areas. At this site, the FSU military repaired vehicles, fueled them, and stored fuel and oil. As a result of these activities, the soil and groundwater throughout the site are highly contaminated, and the contaminants are known to have reached the drinking water wells of nearby residences. This discovery resulted in the intervention of the Hungarian Ministry for Environment and Regional Policy (MERP).

Preliminary Site Evaluation: The MERP conducted preliminary site characterizations in 1991 to define and assess the extent of contamination of groundwater and soil at both sites. Soil and groundwater samples were taken as part of the initial site investigation. All samples were analyzed for total hydrocarbons, polycyclic aromatic hydrocarbons (PAHs), phenol, and benzopyrene. Selected samples were also analyzed for Cr, Ni, Cu, Zn, Cd, Hg, Pb, and As. The results of this investigation identified several specific areas within these two sites as needing further study.

At the Arpad Army Post, 45,700 m^3 (representing a surface area of 25,500 m^2, to an average depth of 2 m) of hydrocarbon-contaminated soil are associated with three main pollutant sources. Hydrocarbon concentrations range from 1,000 to 8,000 mg/kg. The total mass of hydrocarbons contained in the soils is estimated to be 82,250 kg. An estimated 350 kg of free product are floating on top of the aquifer.

At the Monostori Fort, 29,400 m^3 (representing a surface area of 10,700 m^2, to an average depth of 2.8 m) of hydrocarbon-contaminated soil are associated with two pollutant sources. Hydrocarbon concentrations are similar to those found at the Arpad Army Post and the total mass of hydrocarbons contained in the soils is estimated to be 53,000 kg. The dissolved hydrocarbon concentration in the groundwater ranges from 1 to 10 mg/l. An estimated 10,900 kg of free product are floating on top of the aquifer. The human health hazards associated with these contaminated sites include the migration of contaminants off-site and the resultant contamination of drinking water wells, the Danube River, and its associated drinking water function.

Komárom Base Site Information in EnviroTRADE: The Komárom base site information has been incorporated into EnviroTRADE. There is a general overview of the site, as well as detailed descriptions of both Monostori Fort and the Arpad Army Post included in the information system. Site maps, drawings of cross-section hydrogeology, and photographs are also included with the text descriptions. Technical contacts are listed, and important elements have been extracted and coded with the profiles.

The Environmental Research Laboratory (ERL) at the TUB has been involved in environmental research and development, and two technologies under development were added to EnviroTRADE. The first characterizes in situ hydrocarbon contamination in soils, and the second characterizes and monitors air pollution. Both technologies are organized under a general description by the TUB, and technical contacts are provided in all of the profiles. The TUB technologies are among several worldwide technologies identified by EnviroTRADE for potential application at the Komárom Base site.

Additional information from the Komárom base site will be added to the EnviroTRADE system. These include the relative ranking of subareas with regard to cleanup priority and the various approaches selected for cleanup. A critical next step will be the development and input of a quantitative evaluation of the potential risks posed by these sites. This will include estimates of exposure concentrations for potential receptors (human and nonhuman) in all relevant media, as well as the selection of exposure scenarios. Upon completion, the Komárom base site profiles in the EnviroTRADE system will contain a comprehensive collection of site-specific text information, maps, photographs, drawings, and possibly video pictures.

TUB Workstation and Current Database: The EnviroTRADE demonstration system was on loan from SUN Micro Systems. A permanent system installation is currently subject to USDOE evaluation and approval, and will depend also on the availability of alternative funding sources. Proposed plans include the installation of a central region node at the TUB with subnodes at other institutes within the countries of the Central and Eastern European region, depending on expressed needs and interests. Although considerable interest was expressed at the NATO workshop in using EnviroTRADE to assist with the cleanup of FSU military sites in the region, further review of the requirements and funding actions are pending.

Evaluation of Remediation Requirements

Cleanup Requirements: The Monostori Fort is owned by the Hungarian government. The site is intended for use as an historical monument, museum, hotel, and meeting facility to be operated by the City of Komárom. The Arpad Army Post is jointly owned by the local government of Komárom, an association of private property owners, and a number of private companies. The future uses of this site are under discussion.

During 1992 and 1993, some remedial measures were taken at the sites. The last leaking underground storage tanks were removed. At two areas, where the hydrocarbon contamination was the most significant, 5 pumping wells were established. During almost a two month period, 7,500 m^3 of groundwater were treated using a phase separation technique. Some 1.7 m^3 of free product (oil) were removed. The first phase of the contamination isolation effort was finished in mid-December 1992. In 1993, the second phase of the work was implemented when another 2.24 m^3 of groundwater were pumped out of 10 wells in the Monostori Fort. Some 7.4 m^3 of oily emulsion were removed.

The planned actions following the preliminary assessment include: (1) conducting a full assessment, including sampling, analytical evaluation, risk assessment and prioritization, and cost evaluation; (2) performing groundwater decontamination activities; (3) cleaning up contaminated soil; and (4) evaluating results. Unfortunately, a lack of financial resources has prevented further remedial activities, despite the extent and nature of the contamination.

For the phases described above, in addition to funds, technologies are needed which allow the appropriate degree of characterization and remediation at minimal cost. Therefore, following the assessment phase, a detailed analysis of traditional and innovative technologies will be carried out. Different areas of the Komárom Base site will be addressed (for demonstration purposes) using different technologies in order to evaluate and compare their effectiveness. In situ technologies will be preferred (e.g., groundwater pumping and treatment, bioremediation, and vapor extraction technologies). The evaluation of results to be conducted during the last phase will include both the effectiveness of the methods used and a cost and risk evaluation.

The Komárom Base site could be used as a field laboratory for testing and evaluating low-cost technologies which could then be transferred throughout the region for further application. In selecting these candidate technologies, the EnviroTRADE system can be used as a clearing house for the worldwide exchange of environmental restoration information, helping to match technologies with site problems and criteria.

Conclusions

There are reported some 170 abandoned sites similar to Komárom in Hungary alone, and perhaps hundreds of others in the Central and Eastern Europe region that might require cleanup in the near future, depending on the needs of the land and the longer term interests of the country involved. The successful demonstration of EnviroTRADE's capability and the documentation of its advantages and utility in the NATO report may have a significant impact on its potential application throughout NATO, in the Central and Eastern European region, and elsewhere in the future.

The modular architecture of the system allows additional decision support tools to be implemented in the future. Such tools could include: (1) ranking appropriate technologies at a site based on user-specified criteria; (2) analyzing and view flow and transport processes at sites; and (3) calculating the health effects of potential exposures to pollutants at the site.

EnviroTRADE's ability to match the needs of a site with a technology will make it an invaluable tool in remediation efforts worldwide. Using the system, waste site managers with similar environmental problems will be able to identify each other and share successes and failures in applying different technological solutions. In addition, technology developers will be able to identify areas needing new technologies as well as areas in which existing technologies need to be improved.

References

[1]   M. W. Harrington and C. P. Harlan, *EnviroTRADE: An Information System for Providing Data on Environmental Technologies and Needs Worldwide*, SAND92-1525C, Sandia National Laboratories, Albuquerque, New Mexico (1992).

[2]   M. W. Harrington and C. P. Harlan, *Environmental Remediation and Waste Management Information Systems*, SAND93-3987C, Sandia National Laboratories, Albuquerque, New Mexico (1993).

[3]   Richter, Peter I., *Cleanup of an Abandoned Former Soviet Union Military Base in Komárom, Hungary*, (Project proposed for the NATO/CCMS Program), Department of Atomic Physics, Environmental Research Laboratory, Technical University of Budapest, Hungary (1994).

HUMAN HEALTH AND ENVIRONMENTAL RISKS ASSOCIATED

WITH CONTAMINATED MILITARY SITES

Christopher M. Teaf
Center for Biomedical & Toxicological Research and
        Hazardous Waste Management
Florida State University
2035 East Paul Dirac Drive
Suite 226, Morgan Building
Tallahassee, Florida 32310-3700
USA

I. Introduction And Historical Perspective

Public health interest in the investigation, evaluation and remediation of abandoned Soviet military bases in Central and Eastern Europe has intensified as the ownership, and hence the future use, of these facilities has come under the control of the host country in which the base is located. Some of the sites are located in or near heavily populated areas, and in some cases the intention is to convert the base and its associated infrastructure to other uses, including residential housing or commercial facilities.

The abandoned bases vary considerably in size, in the breadth of activities which were conducted on-site and in the duration of use for military operations. As a result, the potential suite of environmental problems which may be encountered is site-specific. While the soils, groundwater and surface water bodies at many of the installations are known to have been affected by surface and subsurface petroleum product contamination, in other instances the materials which may have been disposed on site are not well-characterized, and adequate records often are not available. The larger facilities (e.g., the Komarom base located on the Danube River north of Budapest, Hungary) historically housed hundreds or, in some cases, thousands of troops, often for many years. Functionally, these bases may be viewed as small cities or towns with regard to the municipal-commercial-industrial mix of potential wastes. Potential waste types that may be expected at facilities of this type include:

NATO ASI Series, Partnership Sub-Series, 2. Environment – Vol. 1
Clean-up of Former Soviet Military Installations
Edited by R. C. Herndon et al.
© Springer-Verlag Berlin Heidelberg 1995

- Agricultural materials
- Aircraft fuels
- Cleaning solvents
- Explosives and munitions
- Insulation materials
- Laboratory reagents
- Lubricating and cutting oils
- Medical wastes
- Motor fuels
- Municipal-type solid wastes
- Paints
- Pesticides

Preliminary investigation has been accomplished at many bases. Initial prioritization efforts have focused attention on those which were judged to represent the most significant potential hazards. These preliminary evaluations often have concentrated on identification of such phenomena as fuel releases at vehicle maintenance facilities, which certainly represent areas that may require attention, and which are characterized by readily available methods. Other areas of equal or greater environmental concern may be identified as additional investigations are performed.

A number of approaches have been applied to address the concerns that were identified during the early investigations. In some cases, a "no-action" or "monitoring only" approach has been applied following the investigation process. In other cases, the preliminary remedial efforts have been guided by establishment of the analytical detection limits, or background concentrations, as desirable cleanup targets for groundwater or for soils. This "default" approach may be more restrictive than necessary, and for some contaminant classes (e.g., petroleum hydrocarbons) numerical criteria may have been established by the regulatory agencies responsible for oversight of the remediation. Though these numerical criteria have the effect of limiting the cleanup, their technical documentation often is not available.

In the last few years, it has become increasingly common for toxicologists to employ site-specific assumptions concerning present or future exposure potential, in order to calculate risk-based cleanup concentrations which satisfy the principal requirement of protecting public health and the environment, while keeping remedial costs to a minimum. These risk-based cleanup targets become an increasingly important consideration as analytical chemists improve the detection limits to concentrations of parts per billion (ppb; $ug\ kg^{-1}$ or $ug\ L^{-1}$), parts per trillion (ppt; $ng\ kg^{-1}$ or $ng\ L^{-1}$), or better, for many organic and inorganic analytes.

Risk-based target cleanup concentrations reduce the costs for remediation, while achieving conditions which are adequately protective for humans, wildlife and the environment.

The principles of risk evaluation which are commonly applied at U.S. sites are applicable to facilities in other countries as well, whether or not specific laws are in place to drive the cleanup activities. The risk assessment process can be used in a systematic manner in order to develop target concentrations against which to evaluate the potential risks to public health. In this way, the "higher risk" sites can be assigned a higher cleanup priority, which helps to ensure that financial or technical resources are used effectively. Since the mid-1980's, the United States Environmental Protection Agency (U.S. EPA) has published a number of guidance manuals which summarize approaches and issues to be considered in toxicological risk assessments. These documents, and similar guidance prepared by agencies or groups in Europe, describe methods for site-specific risk evaluation. These guidance manuals and other supplementary information include the following:

- Superfund Public Health Evaluation Manual;

- Alternate Concentration Limit Guidance; Part I: ACL Policy and Information Requirements;

- Superfund Exposure Assessment Manual;

- Workbook of Screening Techniques for Assessing Impacts of Toxic Air Pollutants;

- Risk Assessment Guidance for Superfund (Volume I): Human Health Evaluation Manual;

- Risk Assessment Guidance for Superfund (Volume II): Environmental Evaluation Manual;

- Exposure Factors Handbook; and,

- Supplemental U.S. EPA Risk Assessment Guidance.

The requirements which guide the performance of a site-specific risk assessment depend upon the characteristics of the site and the applicable regulatory requirements of the state or country in which the site is located. However, risk assessment procedures for sites such as the military bases of interest in this paper encompass the following areas:

- selection of indicator analytes, from among those identified in the site characterization activities, that are representative of the toxicity and environmental behavior of the suite of site contaminants;

- estimation of exposure point concentrations of the analytes which are appropriate for the potential receptors (e.g., humans, aquatic species);

- estimation of contaminant intakes for these potential receptors, based on plausible present and future use conditions of the site;

- assessment of toxicity and summary of appropriate regulatory guidance for the identified indicator analytes;

- identification of acceptable toxicant intakes for indicator analytes (i.e., the Reference Doses (RfDs) for noncarcinogens or Risk-specific Doses (RSDs) for potential carcinogens) that are designed to be without appreciable risk of adverse effects to potential receptors;

- comparison between detected site-specific contaminant concentrations with regulatory guidance values, if applicable; and,

- site-specific risk characterization, which incorporates the toxicity assessment and the exposure assessment into a quantitative estimate of potential risks under specified assumptions or exposure scenarios.

II. Analytical Data: Identifying Site Contaminants

The analytes which often are detected in various environmental media at military sites may include polycyclic (or polynuclear) aromatic hydrocarbons (PAHs or PNAs: e.g., naphthalene, chrysene), volatile organic aromatic hydrocarbons (VOAs: e.g., xylenes, benzene, toluene), semivolatile organic hydrocarbons (e.g., phenols, phthalate esters), metals (e.g., lead, mercury, cadmium), other inorganic (e.g., cyanides, sulfur derivatives) and organic analytes (e.g., furans, terpenes, solvents), and pesticides/herbicides. The following is a representative list of items that may be of interest in the investigation of abandoned military bases:

- Acids and bases
- Asbestos
- Drugs, medical supplies
- Fertilizers
- Infectious/biological materials
- Inorganic materials (e.g., cyanides)
- Metals
- Munitions
- Paints
- Pesticides
- Petroleum hydrocarbons
    - volatiles
    - polycylic aromatics
- Radioactive materials
- Solvents (chlorinated, nonchlorinated)

The relative public health risks that are posed by each analyte class will be dependent on the site-specific constituents that are present, their concentration and distribution, as well as the frequency and the magnitude of the potential human exposures that may be relevant to the site.

A major regulatory issue at potentially contaminated sites is the carcinogenic potential of various chemicals (e.g., benzo(a)pyrene, benzene, arsenic). The distinction between potentially carcinogenic chemicals and non-carcinogenic ones should especially be noted, since this distinction forms the basis for critical site-specific decisions regarding potential risks, and will influence the remediation recommendations. Within some classes of chemicals, such as the PAHs, there are potentially carcinogenic (e.g., benzo(a)pyrene) and non-carcinogenic (e.g., naphthalene) members. This distinction, as well as the distinction between aromatic (ring-structured) and aliphatic (non-ringed), have specific applicability to abandoned military bases. At some facilities a total petroleum hydrocarbon (TPH) target concentration (e.g., 100 mg/kg TPH) has been applied to guide remedial activities. However, the composition of the site-specific hydrocarbon mix will determine whether such a generic value is adequately protective. If the analytes are composed primarily of high molecular weight aliphatic compounds, such as may be found in many heavy oils, a 100 mg/kg target may be unnecessarily restrictive. On the other hand, if the composition is weighted toward the potentially carcinogenic PAHs or contains a large concentration of benzene, then the 100 mg/kg target may not be sufficiently conservative. Similar problems may be encountered in cases where, for example, "total volatiles" or "total chlorinated solvent" or "total pesticides" targets are established, since these classes of chemicals may vary widely in toxic potential both between and within categories.

In addition to chemical class and carcinogenicity classification, the environmental fate and transport characteristics of detected analytes should be considered in an evaluation of potential public health risks. As an example, many of the higher molecular weight PAHs (e.g., >5 aromatic rings), as well as the organochlorine pesticides, exhibit a strong affinity for organic or inorganic soil components, coupled with an extremely limited water solubility and low volatility. Thus, leachability typically is very low in the vadose (unsaturated) zone, but may occur in situations where the subsurface soils are in direct contact with groundwater on a continuous or intermittent basis. These considerations make the development of a comprehensive integrated, site-specific sampling strategy for adequate site characterization very important. Possible interactions among environmental media during transport, as well as media that are typically of interest from a public health perspective during risk assessment activities are as follows:

Potential Transport Pathways
- Soils-to-Groundwater
- Groundwater-to-Surface Water
- Water-to-Sediments
- Soils-to-Air
- Water-to-Air

Potential Exposure Media
- Air
- Soil
- Sediments
- Surface Water
- Groundwater

Site characterization involves selecting soil, surface water and sediment sampling and monitoring well installation locations to reflect, as accurately as possible, what is known of the historical operating site activities. The sampling plan should be designed to determine where contamination is not likely, as well as where contamination is expected. Background data are critical for the risk assessment procedure, and small numbers of background samples may increase the probability of a false negative or a false positive error in the decision-making process. A primary objective of the site characterization is identification of source materials (e.g., fuel releases, on-site land disposal areas) and delineation of associated affected soils and water. The identification of source materials provides the opportunity to perform initial remedial actions (e.g., free product recovery) prior to the development of recommended soil or groundwater target concentrations that may guide further remediation. This action permits the most effective removal of the contaminant mass, as well as minimizing its continued distribution throughout the site. In developing the sampling strategy, the early and continued involvement of the toxicologist/risk assessment specialist is important to ensure that the necessary data for the subsequent risk analyses and remedial recommendations will be available.

An ecological risk assessment component may be required in the risk assessment document to address exposure to non-human receptors. Site sampling and data collection activities can be planned to support both an environmental evaluation and a human health evaluation, as appropriate. For example, if sampling of fish or other aquatic organisms is done properly, these data can be used in assessing human health risks from ingestion of fish and shellfish and in assessing impacts to, and potential effects on, the aquatic ecosystem. Specific methods and limitations of ecological risk assessment are undergoing development and rapid evolution in the United States and in European countries.

III. The Risk Assessment Process

A. Introduction

The explicit requirements which guide the performance of a site-specific risk assessment will vary according to the requirements of the controlling environmental regulatory agency (e.g., local, federal, regional). Nevertheless, many of the conventions regarding assumptions and procedures have become consistent across agencies and across programs. Figure 1 summarizes the activities which precede, accompany or follow the performance of the risk assessment process regardless of the entity to which the assessment is submitted. Indeed, the Site Characterization may occur months or years prior to a decision regarding whether a risk assessment is appropriate. The data collection effort forms the basis of the subsequent components of the risk assessment process, which may, in turn, directly affect cleanup decisions.

A risk assessment, as described in the following sections, is composed of the following four primary components:
- Selection of Indicator Analytes;
- Exposure Assessment;
- Toxicity Assessment; and,
- Risk Characterization.

B. Selection of Indicator Analytes

This process typically employs a concentration/toxicity screening approach to identify analytes which comprise the most significant proportion of the potential risks in each environmental medium (e.g., groundwater, soils, surface water, sediments). Based on the detected soil, sediment, surface water or groundwater concentrations and, to the extent possible, the established toxicity benchmarks (e.g., Reference Dose (RfD) or Carcinogenic Slope Factor (CSF)), indicator analytes are identified which individually represent a significant proportion of the potential risk. This selection step ensures that the risk assessment will emphasize the analytes which are the principal contributors to potential site hazards.

Generally, if less than 10-15 analytes are detected at a site, it is appropriate to select all of them unless a peculiar situation exists (e.g., a case where one analyte represents greater than 95% of the aggregate risk). In most other circumstances (e.g., sites with greater than 10-15 detected compounds), the concentration/toxicity screening approach is a useful method by which to narrow the list of substances that will be considered in the risk analysis. Under some

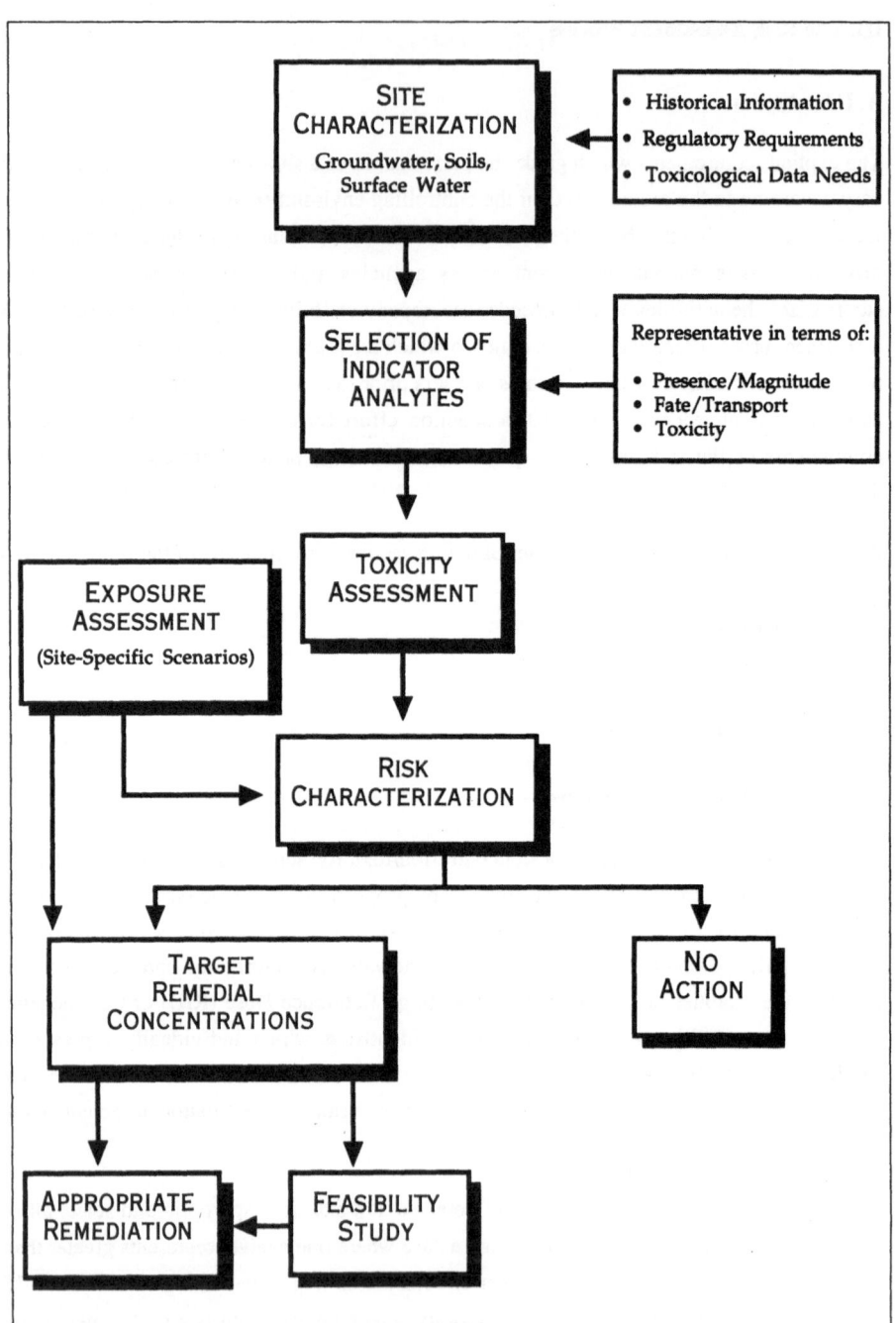

FIGURE 1

programs, regulatory agencies may recommend that "all hazardous constituents" at a site be included as indicator analytes in the risk assessment, as opposed to allowing the above selection process. As a practical matter, chemicals are targeted for initial evaluation if detected at the site in concentrations which are in excess of site or regional background concentrations (e.g., concentration greater than 2 times site or regional background), or are present in groundwater in excess of local drinking water standards.

C. The Exposure Assessment

The development of selected site-specific exposure scenarios for risk calculations is conducted to establish the frequency, duration and magnitude of exposures which may be expected for individuals either on-site or off-site, under present-use and projected future-use scenarios. As such, the exposure assessment activities to a great extent will determine the outcome of the risk analysis. Two distinct categories of risk assessment presently are recognized:

- *Deterministic* - based upon calculations that employ discrete numerical assumptions which represent default values or which are judged to represent a selected segment of the population distribution (e.g., 95 percentile) for each input variable. This is the simplest approach, and allows manipulation of the individual variables. However, given a set of specified input parameters, this process allows modification only of the target concentration in the establishment of remedial options.

- *Probabilistic* - based upon statistical evaluation of the frequency distribution for individual variables. If a desired result is, for example, to achieve conditions which represent an aggregate risk equal to $p < 0.05$ or $p < 0.01$ (alternately stated as 95% Upper Confidence Limit or 99% Upper Confidence Limit, this process allows for the input parameters to be altered independently of one another in iterative fashion to produce different combinations of exposure conditions that will satisfy the end objective. Thus, exposure controls as well as target concentrations may be combined feasibly.

In practice, a number of discrete exposure assumptions have become standard in risk evaluations (e.g., adult or child body weight, water consumption rate, inhalation rate, length of human lifetime), and many environmental regulatory agencies in the United States have established explicit requirements for use of certain parameters. One difficulty with such "deterministic" approaches is that the sequential use of conservative values for a number of input parameters often results in an extraordinarily conservative final result.

Historically, the deterministic approach has been almost universally applied. Because the risk assessment often is used in negotiations regarding what may or may not be appropriate for the future use of the site, the risk assessment must utilize reasonable site-specific assumptions, not only those "worst case" or "default" scenarios that may overemphasize site risks. In the last few years, probabilistic risk assessment techniques (e.g., Monte Carlo simulations) have been used to estimate the statistical distributions for some of these parameters. This may represent a useful alternative to what have become the standard calculations. If their utility is validated, such approaches may see increasing application as an alternative to deterministic approaches.

D. Toxicity Assessment

This activity results in the collection, summary and presentation of information on the relevant chemical, physical and toxicological characteristics of the selected indicator analytes and, if available, the regulatory guidance values. Many commonly encountered analytes have U.S. EPA-derived RfD values or CSF values available, and typically those regulatory benchmarks will be useful in the risk analysis calculations. In some cases, however, an analyte which makes a significant contribution to potential site risks (i.e., widespread detection at high concentrations) may not have an RfD or CSF available. For these analytes, the toxicologist must review the published literature in order to provide a reasonable profile for an estimate of toxicity, which is then used in the risk characterization process. The information collection process described in this section typically results in preparation of a toxicological profile that provides an appropriately referenced source for documentation purposes.

E. Risk Characterization

This component of the process combines the results of the Exposure Assessment and the Toxicity Assessment in order to estimate in quantitative fashion the potential noncarcinogenic or carcinogenic risks under assumed conditions for the specific exposure scenarios. Non-carcinogenic risks may be expressed as a fraction of the presumed acceptable dose or benchmark value (e.g., RfD). This fraction or ratio, often termed the Hazard Index or Hazard Quotient, represents a potentially unacceptable circumstance at chronic values greater than unity (1.0), that is, where the estimated intake exceeds the benchmark dose. For potential carcinogenic risks, guidelines in the United States historically have identified an acceptable risk range which extends from 10^{-4} to 10^{-7} (or 1:10,000 to 1:10,000,000) for estimated lifetime excess cancer risk under different regulatory applications. In the U.S. and elsewhere, most agencies have narrowed the acceptable window to the range of 10^{-5} to 10^{-6}.

Remediation usually is not required beyond that which is required to achieve 10^{-6}, and cleanup strategies are usually not selected which cannot achieve 10^{-4} calculated excess lifetime cancer risk. Exceptions to these guidelines have occurred at some U.S. sites.

The flexibility which may be incorporated into a given site-specific risk assessment process is a function of the judicious development of the exposure scenarios and the ability to defend, on a toxicological basis, the assumptions that are associated with these scenarios.

IV. Development Of Risk-Based Remedial Targets

As shown in Figure 1, the Risk Characterization process may lead to a conclusion that potential risks under existing on-site conditions are within acceptable limits and that remediation is not required (the "No Action" alternative). Alternatively, target concentrations may be proposed for one or more environmental media, in order to guide appropriate remedial activities. Such values, often termed Site Rehabilitation Levels (SRLs) or Target Concentration Limits (TCLs), are developed from the estimated risks associated with site-specific exposure scenarios. This process involves factors such as potential additive toxicity among analytes, possibility of simultaneous exposure by multiple pathways (e.g., oral and dermal) and multiple environmental media (e.g., groundwater and soils).

The proposed remedial target concentrations exert a strong influence on the technical and economic feasibility of the site cleanup. Figure 2 illustrates that for a certain range of remedial concentration targets, costs may be affected in only a limited way. At some point, however, the curve steepens and it becomes incrementally much more expensive to recover each mass unit of contaminant. Thus, the remedial targets must be established in light of all relevant considerations. As an example, in the absence of proposals to the contrary, soil concentrations in the 1-10 mg/kg (ppm) range may be required for some potentially carcinogenic PAHs, such as may be found in some tars or unrefined petroleum products. However, in some cases, soil concentrations for aggregate PAHs as great as 700 mg/kg have been deemed acceptable at PAH-contaminated sites, if site access is limited or exposure can be otherwise prevented with reasonable certainty. This decision may hinge primarily on the site characterization presented in the risk assessment, and is strongly dependent on site-specific considerations such as the planned future use, restrictions of use for the site and which PAHs are present at the site. Such considerations are of particular interest in addressing the question of appropriate remediation for abandoned Soviet military bases in Central and Eastern Europe.

In the development of the TCL/SRL values, consideration also should be given to the existing standards or guidelines for the environmental medium of interest (e.g., Drinking Water

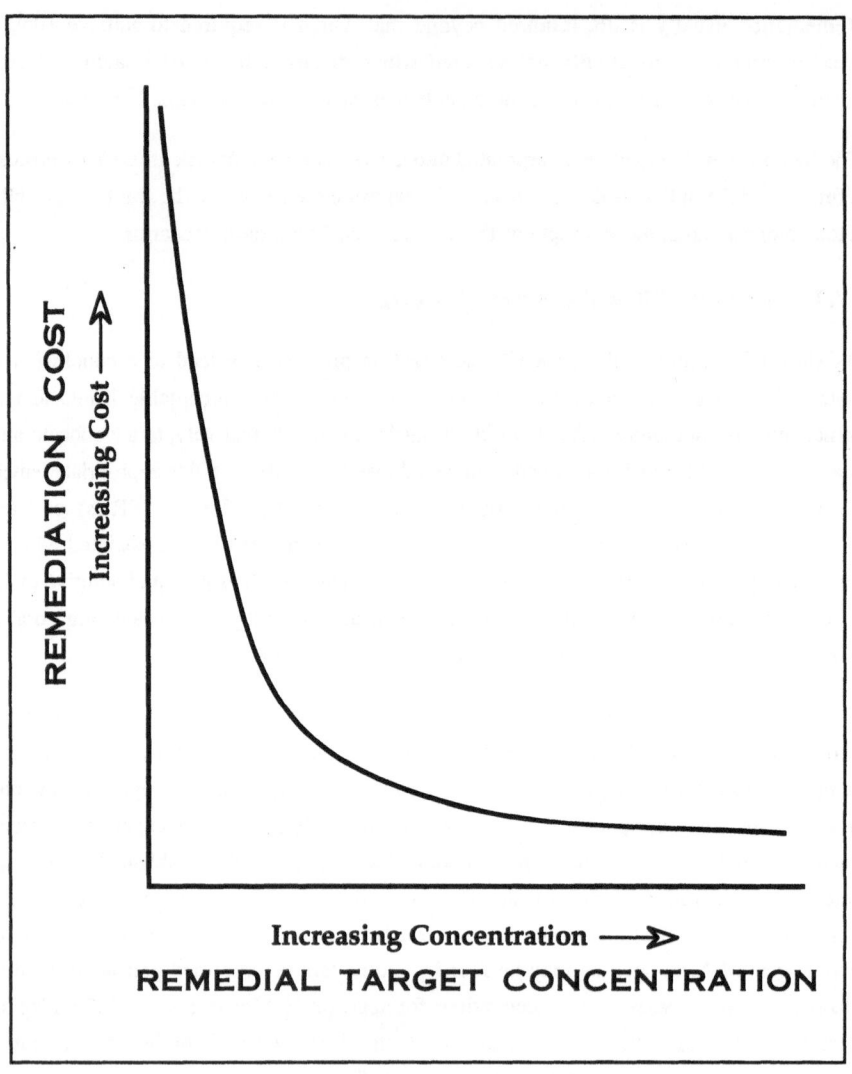

FIGURE 2

Standards). This facilitates the presentation of site-specific data which may support a decision, for example, not to default to a general regulatory value. As an example, at sites overlying strongly saline aquifers, or those which exhibit ambient or natural concentrations of dissolved analytes which render a water source nonpotable (e.g., iron, copper, manganese, total dissolved solids), the application of drinking water standards may not be appropriate. In these cases, it is necessary to propose and document alternative remedial target concentrations based on site-specific exposure potential, rather than on general regulatory guidelines. As noted previously, the extent to which such proposals are accepted at any site will reflect the technical defensibility of the proposal and the supporting documentation, represented by the Risk Assessment report.

Another area of emerging interest is the potential applicability of a "deed restriction" for the property in question. Depending upon local zoning criteria and land use, such "restrictive covenants" provide a means to preclude or to limit future activities that may be conducted on-site. This, in turn, limits the applicable exposure scenarios and influences the proposed TCL/SRL values. Such an approach may be of particular interest at sites (e.g., converted military installations) for which future uses may be more accurately projected and/or controlled.

V. Summary And Conclusions

The evaluation and remediation of contamination which is detected or suspected at abandoned Soviet military bases represents an enormous financial and technical investment of resources. It is essential that these resources be used in a way which adequately protects human health and environmental considerations in the context of technical, regulatory and economic considerations (Figure 3). The risk assessment process provides a systematic approach by which to evaluate site hazards and to establish target concentrations for the effective application of remedial alternatives to soils or groundwater. These alternatives range from "No-Action" to standard intrusive methods such as comprehensive soil excavation and groundwater recovery/treatments. The establishment and implementation of reasonable, site-specific remedial guidance concentrations and appropriate remedial alternatives depend on the quality of the risk assessment documentation. Without such detailed justification and supporting calculations, the position of the environmental agencies may be to require remediation to "background" conditions or to analytical detection limits, which may not be necessary from a public health perspective, and which may exhaust the limited financial and technical resources that can be brought to bear on these and other problem sites.

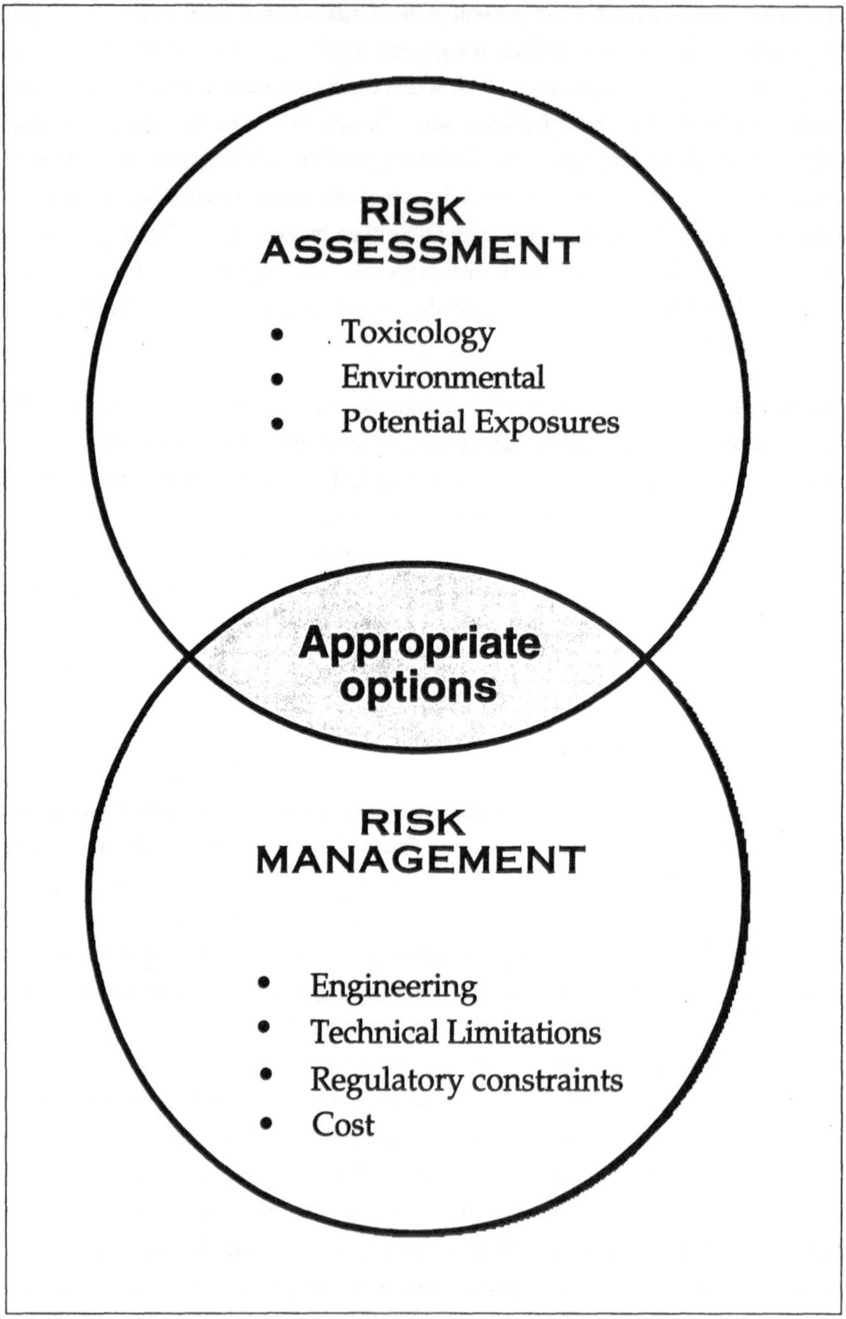

RISK ASSESSMENT

- Toxicology
- Environmental
- Potential Exposures

Appropriate options

RISK MANAGEMENT

- Engineering
- Technical Limitations
- Regulatory constraints
- Cost

FIGURE 3

SELECTION OF REMEDIAL OPTIONS FOR CONTAMINATED SITES

H.J. van Veen and A. Weenk
TNO Institute of Environmental and Energy Technology
P.O. Box 342
7300 AH Apeldoorn
The Netherlands

Summary

This paper presents a systematic approach for prioritizing options for the remediation of contaminated sites. This approach is based on a step-by-step plan to solve environmental problems called STEPS.

1. Introduction

Countries are facing many environmental problems and limited budgets to solve these problems. Thus, priorities have to be set and choices have to be made. To an increasing degree, there is a need to base these priorities and choices on rational objective criteria so that the environmental funds are spent as efficiently as possible.

Environmental decision-making is often very difficult because of the confusion about the meaning and value of concepts such as problems, causes, consequences, effects, targets, measures, solutions, etc. The inconsistent use of these words considerably hampers the discussion. Other obstacles can be the lack of competent authority and complex legislation.

This paper describes a general systematic approach for evaluating environmental problems called STEPS. Included in this paper is a specification for contaminated sites.

2. STEPS

STEPS (Systematically Tackling Environmental Problem Solving) is a step-by-step structured approach for environmental problems and solutions. The plan generally consists of three components:
- identifying and prioritizing environmental problems;
- generating improvement options (actions);

NATO ASI Series, Partnership Sub-Series, 2. Environment – Vol. 1
Clean-up of Former Soviet Military Installations
Edited by R. C. Herndon et al.
© Springer-Verlag Berlin Heidelberg 1995

- assessing the generated improvement options on environmental merit and achievablility, and in accordance with sustainability principles.

Out of these three components an overall judgment is made concerning the desirability of the improvement options. These components are visualized in Figure 1.

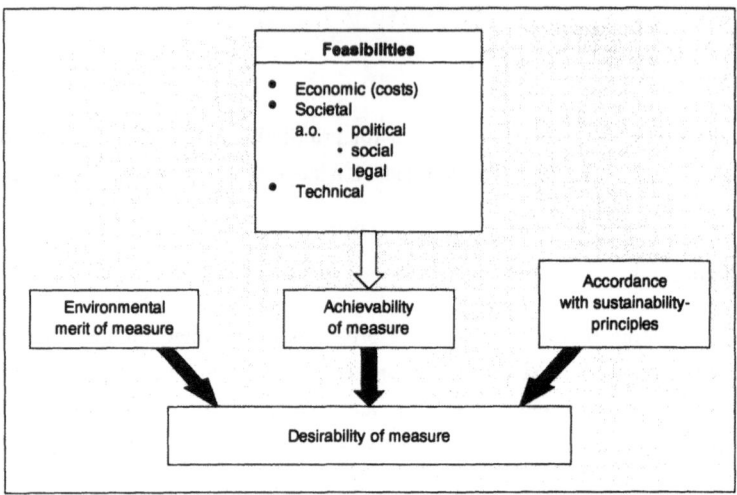

Figure 1   STEPS-structure

STEPS is divided into 5 blocks (A-E), as shown in Scheme 1. After performing the steps, a sensitivity analysis can point out the influence of various uncertainties. Next, desirable improvement options can be selected for implementation.

Block A: Problem analysis

In Block A, environmental problems of involved parties are identified and ranked.

Problem domain
1. Mark out the problem domain.

Environmental problems
2. Determine environmental problems.
3. Rank these bottle-necks on importance.
4. Define targets with respect to the problems.
5. Trace the causes of the environmental problems.

Block B: Improvement options and their environmental merit

In Block B, improvement options (measures) are generated. A preselection is performed, based on known achievabilities and the environmental consequences of options.

6. Generate improvement-options.
7. Make a rough preselection based on known achievabilities.
8. Determine main and side-effects of the improvement-options.
9. Determine the environmental merit of the improvement-options.
10. Make a rough preselection based on the environmental merit.

Block C: The achievability of improvement-options

In Block C, the goal is to determine to what extent the preselected improvement options can be realized with the intended environmental merit.

11. Determine the technical feasibility.
12. Determine the (micro- and macro-) economical feasibility and the environmental efficiency.
13. Determine the social, political and legal feasibility.

Block D: Sustainability-principles

In block D, the accordance of improvement-options with principles, which are generally thought to apply in a sustainable society, is estimated.

14. Subject the improvement-options to sustainability-principles.

Block E: The desirability of improvement-options

In block E, the results of the previous blocks are brought together. A final assessment results in the desirability of the improvement-options

15. Determine the desirability of improvement-options, based on their environmental merit, (from block B) their achievability (from block C) and their accordance with sustainability-principles (from block D).

Scheme 1

2.1 Some important concepts in STEPS

Cause-effect chain

In block A, problems are stated as undesired effects or risks to human health and the environment. These problems arise from a change in environmental quality, which is the result of human activities. One can say that activities are performed to fulfill human needs. These elements can be represented in a general cause-effect chain:

human needs

⇓

human activities

⇓

use of the environment

⇓

change in environmental quality

⇓

effects and risks to human health and the environment

When analyzing a problem, the general cause-effect chain is specified for a problem situation. This specification gives insight into the nature of the problem and possible approaches for application of improvement options.

Environmental merit

The environmental merit of improvement options (step 9) can be determined by using various principles and tools, including Life Cycle Analysis and Assessment (LCA). LCA is becoming an internationally accepted method. By using LCA, aggregation techniques for emissions and use of raw materials, along with a multi-criteria analysis technique, a clear and accepted method for determining the environmental merit of improvement options can be developed.

Achievability

As seen in the STEPS structure, the achievability of improvement options (Block C) depends on several "feasibilities" (i.e., economic, societal and technical). The achievability of improvement options has been, to this point, kept strictly apart from the environmental merit.

The goal of this has been to keep discussions about environmental aspects pure and separated from discussion on implementation aspects.

Sustainability principles

A third factor which influences the desirability of an improvement option is the accordance with sustainability principles. This factor reflects ideas related to laws and guidelines in a sustainable society. Examples are:
- closed life cycles of material;
- polluters are made responsible;
- improvement options are cause-directed (prevention, source-oriented, process-integration versus end-of-pipe); and
- improvement options lead to involved actors who are inspired to think for themselves about environmental problems and solutions.

3. STEPS for contaminated sites

3.1 Block A: Problem Analysis

Problem domain (1)

This paper is focused on the domain of contaminated sites which can be further specified to, for instance, contaminated military sites in Hungary. Only the options to improve these sites are prioritized; the problem of prioritizing site remedial options together with other environmental problems will not be considered.

Making an inventory and ranking of contaminated sites (2, 3)

In this step, the inventoried sites are ranked according to the urgency for remediation.

In the Netherlands, a two-stage approach to problem-ranking for the domain of contaminated sites is conducted. The first stage is a preliminary "screening" investigation of a site by which the presence or absence of expected contaminants is established. If the preliminary investigation indicates contamination, a further investigation is executed to determine the extent of the pollution and to determined whether serious danger to human health and/or the environment exists.

This further investigation consists of two phases. In the first phase, potential risk is calculated by comparing observed contamination levels with a predetermined concentration level called the "intervention value." If this value is exceeded, there is a legal necessity for remedial action. In phase two the actual (site-specific) risk is calculated. By means of exposure analysis it will be determined whether remedial action is *urgent* or not. The analysis is an additional step-by-step approach as given in Scheme 2 [Berg, R. van den, et al., 1993]:

Scheme 2. Step-by-step actual exposure analysis

1. Identification of exposure pathways;
2. Analysis of main contributing pathways;
3. Additional measurements;
4. Calculation of human exposure;
5. Determining human health risk;
6. Determining environmental risk;
7. Possible future spreading of contamination;
8. Interpretation with special attention to background exposure and toxicology.

According to this Dutch approach (and other similar approaches) it is possible to rank the sites according to the urgency for remedial action based on risk.

Remediation targets (4)

According to the STEPS approach, targets for the remediation have to be defined. Targets can be defined on various levels, such as:

- Restoration of the multifunctionality of the soil. This indicates that the site is completely restored such that all functions of the land (agriculture, residential, etc.) are possible. This is probably one of the highest targets to be defined.
- Reduction of the potential risk. This is meant to avoid any risk to humans or the environment without total restoration.
- Reduction of actual risk. This is commonly used in industry. It means that employees are protected and that spreading of the contaminant is avoided.
- Avoidance human exposure and acceptance of the contamination.

The target to be chosen depends on various aspects, such as:

- The intended function or use of the site after restoration (landuse); and
- The national policy targets.

However, in a systematic approach it is important to define concrete targets.

Source of contamination (5)

The source of soil contamination is mostly a historical use of the land. Often the source is no longer functioning. However, when the source still exists, remediation might not be useful.

3.2 Block B: Improvement options and their environmental merit

Improvement options (6)

Improvement options in site remediation mean generally the application of remedial technology. Numerous technologies have been developed in various countries over the last two decades. Remedial options often consist of a combination of these techniques. The composition of the remedial approach is site-specific. The different target levels (3.1) lead to different remedial approaches.

Target: Restoration of multifunctionality

Examples of technologies:
- Excavation and treatment of the contaminated material by various techniques like thermal desorption or soil washing; or
- *In-situ* treatment of soil and groundwater by leaching or biological techniques.

Target: Reduction of potential risks

Examples of technologies:
- Pumping and treatment of the groundwater; or
- Geohydrological isolation.

Target: Avoiding human exposure

Examples of technologies:
- Covering the site with clean soil; or
- Use of signs, such as: "Trespassers will be prosecuted" and fences.

Depending on the target, improvement options based on knowledge and experience can be generated by experts. In the Netherlands, the RIVM is developing a decision support system which generates remedial approaches based on site information.

Preselection (7)

The preselection is meant to discriminate sense from nonsense. Due to the recently developed technologies, numerous remedial approaches can be generated. It is necessary to eliminate those options which are absolutely not feasible according to the criteria of Block C. This rough preselection should be performed on the basis of an educated guess (i.e., expert evaluation) and should result in a list of relevant remedial options.

Main effects and side effects (8)

The main effect of remedial options is defined by the target of remediation (e.g., clean soil, risk reduction). However, in addition to this main effect, remediation activities can result in a number of environmental side effects such as:
- emissions to the atmosphere;
- emissions to surface water;
- production of waste;
- consumption of energy; and
- consumption of chemicals.

The determination of side effects should not be overdone, for instance the production of raw material for which energy is consumed are side effects of the consumption of chemicals. This energy consumption is a second or third order side effect. Regarding the remediation of contaminated soils, it is proposed that only first order effects be considered.

Environmental merit (9)

The definition of environmental merit is one of the topics of current environmental debates. In the past, environmental actions were often focused on only one component of the environment such as wastewater purification or air pollution abatement. These actions were aimed to clean the water or the air, without regard for side effects like energy-consumption and production of sludge. However, these actions are often judged on more comprehensive environmental criteria: "What is the total impact of the remedial action on the whole environment?".

It is beyond the context of this paper to discuss this issue in detail. It is proposed that a general approach to assess the environmental merit of soil remedial options be deduced from the Life Cycle Assessment (LCA) approach [Heijnings, 1993]. In a LCA, the environmental merit is classified according to a number of classification themes as given in Scheme 3.

Scheme 3

Classification themes	
GW:	Global warming
OD:	Ozone depletion potential
HT:	Human toxicity
AT:	Aquatic toxicity
TT:	Terrestrial toxicity
OC:	Photochemical ozone creation potential
AC:	Acidification potential
NP:	Nutrification potential
Wa:	Landfill space (non toxic waste)
TW:	Confined disposal space (non toxic waste)
En:	Exhausting fossil fuels
Ma:	Exhausting mineral raw materials

In some cases, the contribution of a specific remedial option to some classification themes can be judged as negligible. By doing this explicitly, the analysis remains clear. Using the classification themes as a checklist also prevents the omission of relevant considerations.

The weighing of classification themes is not only a political choice, but also depends on the national or local attitude toward these issues. In the LCA approach, all themes should be weighed equally.

In the LCA, emissions and consumption processes are calculated by the use of a database with information on standard process emissions. These emissions are classified into a theme score according to substance specific classification factors. All theme scores together give an LCA environmental profile (Figure 2).

To calculate the environmental merit of soil remedial options, a similar method can be used. In the Netherlands projects are now in progress for the implementation of assessment elements of the LCA approach for the calculation of the environmental merit of soil and sediment remediation.

Selection of an option according to environmental merit (10)

If the environmental merit is known for all relevant options, it is possible to rank the options according to their respective environmental merits. The option on top of the list is the best option for the environment.

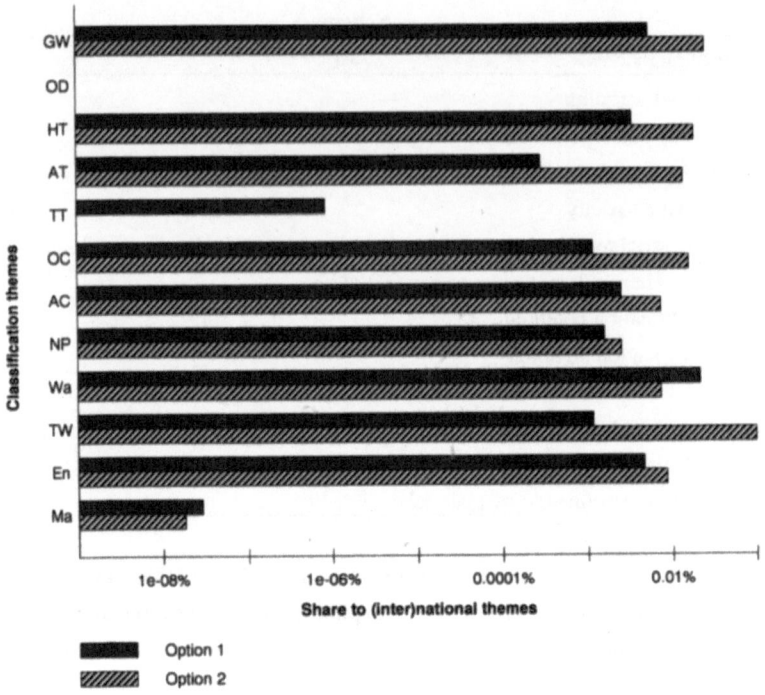

Figure 2   Example of an LCA environmental profile

3.3 Block C: The achievability of improvement options

It is important in STEPS that a distinction is made between environmental merit and achievability of remedial or improvement options.   The goal of this is to not confuse environmental and implementation aspects of an improvement option.

The achievability depends on several feasibilities.   Some of these feasibilities are mentioned here, but local circumstances may introduce other feasibilities.

Technical feasibility (11)

The technical feasibility is based on certain criteria:
- The reliability of a technique;
- The time that is necessary for the execution of an option, which depends on capacity of the applied technology; and
- The state-of-the-art technology to be applied.

Economic feasibility (12)

The economic feasibility represents mainly the costs of the different options, but also the availability of funding, which may change from year-to-year.

Social, political and legal feasibility (13)

The social attitude toward soil contamination differs from country-to-country and site-to-site, but particularly in densely populated areas there is a large social pressure demanding remediation of contaminated sites. This may result in the rejection of some improvement options such as covering (e.g., use of cover material to inhibit infiltration and migration of contaminants).

In some cities in the Netherlands, political statements have influenced the choice for remedial options. This makes some options more feasible than others.

Legal feasibilities are, for instance, related to licenses, permits, etc. Emissions to water or air are restricted, which may result in the rejection of some options. This may also occur when a long period is necessary to obtain permits.

3.4 Block D: Sustainability principles (14)

With respect to sustainability principles the use of soil may be seen as part of a controlled chain in which the soil is restored after use. TNO has presented a concept for sustainable soil use [Palsma, A.J., et al., 1993]. It is beyond the scope of this paper to go into detail about soil sustainability. So far there are hardly any sustainability principles operational to the extent that they can be used for prioritizing improvement options for contaminated soils.

3.5 Block E: The desirability of improvement options (15)

The desirability of improvement options for contaminated sites is a judgment based on environmental merit, achievability, and accordance with sustainability principles. Some decision makers in the Netherlands are charmed by an aggregation of the environmental merit and the costs of the option into one number: the environmental efficiency.

STEPS attempts to present tables of values for environmental merit, achievability and accordance with sustainability principles. Based on this information, remedial options can be prioritized directly.

4. Conclusion

This paper presents a systematic approach for prioritizing options for the remediation of contaminated sites. As such, it is able to harmonize discussions about problems and solutions between parties involved.

Parts of this approach are already implemented, and parts have yet to be developed. This means that the approach is not fully proven, so there might arise much discussion concerning parts of this approach. However, the authors are convinced that a systematic approach is essential for achieving environmentally sound and cost-effective remediation plans for the large number of contaminated sites all over the world. STEPS may provide the backbone of such an internationally accepted approach.

References

Berg, R. van den, et al. (1993). Risk assessment of contaminated soil: Proposal for adjusted toxicologically based Dutch soil clean-up criteria. Contaminated soil '93, 349-364, Kluwer Academic Publishers, The Netherlands.

Heijnings, R. et al. (1993). Environmental life cycle assessment of products. NOVEM report, The Netherlands, ISBN 90-5191-064-9.

Palsma, A.J., et al. (1993). Sustainable soil use, a TNO-concept. TNO-ME, Apeldoorn, the Netherlands.

REMEDIATION STRATEGIES FOR CONTAMINATED (FORMER)

MILITARY SITES

Merten Hinsenveld[*]
Research Professor, University of Cincinnati
Fröbelstrasse 10
69226 Nussloch
Germany

Introduction

The potential cost of contaminated sites remediation at (former) military bases in Eastern and Western Europe is enormous. Most countries can afford the remediation of only a small fraction of their contaminated military sites each year. Consequently, many contaminated (former) military sites will not be remediated in the foreseeable future. In the meantime, contaminants are migrating in the environment with rain- and groundwater movement, continuously expanding the area of contamination and increasing the cost of future remediation.

A large fraction of environmental funding is generally spent on eliminating immediate health threats and on alleviating the most pressing problems of drinking water deterioration. Given the expansive character of the contamination, however, control measures, i.e., measures that limit expansion of existing contamination or prevent new health threats from occurring should sometimes receive higher priority than measures that correct existing problems. The dynamic and multi-faceted nature of the contaminated sites problem, combined with the potentially huge financial consequences of any measures taken (including, paradoxically, the no action alternative), poses a complex and delicate problem that is just beginning to be examined. This paper gathers a number of the most essential elements in a remediation strategy and suggests a direction towards a consistent and flexible remediation program.

---

[*]   Presently, Special Consultant to Headquarters, U.S. Army Europe and 7th Army, Heidelberg, FRG.

NATO ASI Series, Partnership Sub-Series, 2. Environment – Vol. 1
Clean-up of Former Soviet Military Installations
Edited by R. C. Herndon et al.
© Springer-Verlag Berlin Heidelberg 1995

Starting Points for a Remediation Strategy

Key starting points for a remediation strategy in the most general sense are [Hinsenveld and Assink, 1988]: (1) to meet the requirements for human health and the environment in the short term and in the long term, (2) to meet the needs for water resources in the short term and in the long term, (3) to meet the needs of the actual or planned activity in the site in the short term and in the long term, and (4) to meet some criteria of cost-effectiveness in the short term and in the long term. These points are described in greater detail in the following section.

Point 1: to meet the requirements for human health and the environment in the short term and in the long term (risk reduction).

The most obvious, although not the most commonly encountered, reason to begin a remediation effort is to eliminate or mitigate unacceptable health risks to human beings and/or to the environment. What is considered "unacceptable" can vary by country and by organization. For humans, the remediation can aim at reducing the direct exposure of humans to the contaminants, at reducing the exposure through consumption of produce and at reducing the risk of deterioration of drinking water quality for unregulated users (e.g., those in rural areas). For the environment, it aims at reducing the deleterious effect of the contaminants on flora and fauna.

Point 2: to meet the needs for water resources in the short term and in the long term (i.e., resource conservation).

A second reason for remediation, perhaps most common in occurrence, is the protection of groundwater resources. Water is an excellent solvent fro some contaminants in soil. It transports them through the subsurface and into valuable groundwater aquifers, which then no longer can be used as drinking water resources. It often is difficult and costly to remove these contaminants from drinking water. Flowing groundwater, generally, emerges as surface water (e.g., river, lake, sea) and, when it is contaminated, it can also contaminate the receiving surface water body.

Although groundwater wells usually are closed down before a population actually is affected severely, a loss of groundwater resources can have dramatic consequences, and may lead to large expenditures when new extraction areas need to be explored and prepared for drinking water use (requiring the installation of new wells, purification units, and distribution systems). Once contamination is beyond control, deterioration of groundwater aquifers and damage to receiving water bodies can be long lasting and extensive.

Point 3: to meet the needs of the actual and planned activity on the site in the short term and in the long term (i.e., functionality restoration).

Contamination can result in an actual loss of certain uses and sensitive functions of the site (such as its suitability for housing). When the contamination is minor, the mere listing of the site as contaminated can lead to a perceived risk and restrictions on usage, which may have the same functional result as an actual risk. In both cases, listing a site as polluted deprives it of a part of its economic and functional value.

Irrespective of the actual risk or true danger to resources, remediation may be necessary in order to restore the asset value of the property by removing a *perceived* risk. In addition, in some cases, remediation may be necessary to allow the use of the property more "sensitive" function which is prevented by the prevailing contaminant concentrations (e.g., reclamation of industrial land for residential or agricultural purposes).

Point 4: to meet some criteria of cost-effectiveness in the short term and in the long term (i.e., cost reduction).

It is probably unavoidable that human activities at military sites will result in an increase in the concentration of anthropogenic components, typically defined as contaminants, in the soil. A cost-effective strategy addresses two factors: (1) it does not seek to remediate to lower levels than necessary for the specific functions of the site, and, (2) it takes into account that after cleanup, new pollution may occur. In other words, a cost-effective strategy aims at managing and controlling the contamination to minimize long term cost, rather than at conducting an "ultimate cleanup". When a site is turned over to another agency or owner, or new functions of the site are considered, an "ultimate remediation" is often conducted. In this case, the asset value or requirements of interested buyers, rather than cost-effectiveness as a function of intended future land use drives the remediation.

Remediation Strategy

Exposure to contaminants can be quantified on the basis of " Source-Path-Target" model. This model states that, to be of concern, the contaminants must reach a person or resource (target) through a medium (path) that connects it to the site (source).

Remedial actions seek to break the source-path-target chain by eliminating one or more of its components, hence we distinguish source-, path-, and target-related remediation techniques. Each remedial action aims at reducing a risk, protecting a resource, and/or maintaining or restoring a function of the site in the context of controlling overall (ultimate) costs. Different goals of remediation typically lead to different remediation alternatives.

Source-related Remediation Measures (Cleanups)

These measures aim at eliminating or mitigating the source. According to the spatial distribution of the contamination, we distinguish point sources (such as fuel stations, chemicals storage facilities and spills) line sources (such as leaking sewer systems and pipelines) surface sources (such as large hardstands with many local spills) and atmospheric deposition.

Initially, it was thought that it would be simply a matter of developing the proper technology to remove most contaminants from the soil. Presently we know that the restoration of contaminated soil to pristine levels is technically extremely difficult, if not impossible. Nevertheless, exceptions exist, mainly for organic contaminants that may be treated using thermal techniques. Above all, restoring multifunctionality of the site soil often is not affordable and not always necessary, based on health or environmental considerations.

Removal of contaminated soil and storage or deposit at a landfill is an example of a source-related remediation. Lightly contaminated soil can, after excavation, often be used in non-sensitive bulk applications (e.g., road building, asphalt), which makes more sense and is often cheaper than depositing it in a landfill. When the contamination extends below the water table, hydraulic measures and/or physical barriers may be needed to keep the excavation pit dry. For this reason, ongoing source-related cleanup and removal actions are often accompanied by isolation techniques. Cleanup of extracted groundwater may lead to large expenditures.

Free phase oil removal from groundwater as a source-related measure is usually a matter of common sense and cost-effectiveness in the long term. It should not be delayed unless there are strong reasons to do so, since the removal of dissolved oil constituents is orders of magnitude more expensive that removal of the oil in a free phase on a mass basis.

Path-related Measures (Isolation)

Contaminants can be transported via a number of migration paths. These paths include subsurface migration through the vadose zone (soil air), through the saturated zone (soil water), and surface migration through air and dust particles. Isolation of the contamination can be done by installing physical barriers (surface capping, barrier walls) or by installing hydrological barriers (groundwater barriers) that cut off leachate and/or groundwater transport routes. Covering the surface with a 1 meter layer of clean soil, combined with groundwater control measures is a commonly chosen path-related remediation technique in urban areas in the Netherlands.

Similarly, covering an area of contaminated soil for use as a paved parking lot usually eliminates the direct health threat to humans. When contaminants have not reached the groundwater, pavement is an adequate remedial action, since it prevents the leaching of contaminants with rainwater into the groundwater. In general, physical isolation is more expensive than merely limiting access, but is less expensive than cleanup.

Target-related Measures (Access Limitation)

When humans are prevented from contacting the contaminant source through one of the possible exposure routes, the contamination poses no human health risk. Humans can inhale contaminated air, including vapors generated during bathing and showering, as well as inhaling contaminated soil particles. They can drink water, ingest soil particles and consume produce (food). They can also be exposed to contamination by direct contact with the skin (dermal). Target-related remediation measures most commonly are interim measures designed to delay the implementation of a more permanent remedy.

Target-related measures can aim at limiting or prohibiting general access to the site or limiting the functions of the site by declaring an area as industrial and/or not allowing children to play at that site (limit contact with exposure routes). It is clear that resources, being one of the possible targets, cannot be protected in this manner since groundwater aquifers cannot be moved or cordoned off.

The Risk of Soil and Groundwater Contamination

Soil contamination is often primarily a subsurface problem, with limited direct effects to human health and, combined with the limited time of exposure, a site visit does not necessarily lead to a significant risk. *As a rule, the exposure of humans to contamination on an industrial or military site is small, the most important route typically being inhalation or ingestion of contaminated soil particles.*

Risk assessments provide information on the actual potential health affects of a contamination situation, which is related to the exposure situation, the behavior of the contaminant, and the toxicological response of the target species to the contaminant. Any background exposure should be taken into account in the risk analysis. In relation to contaminated sites, two kinds of risk are defined: (1) risk to human beings and (2) risk to the environment.

A risk to humans is unacceptable when the maximum tolerable risk level (MTR) is exceeded. For non-carcinogenic compounds is defined as a level less than or equal to the acceptable daily intake (ADI). The MTR is derived from toxicological considerations, such as NOELs (no observed effect level) and, for carcinogenic compounds, a statistical criterion (i.e., one extra case of cancer in 10^5 people is often considered acceptable or "safe"). Risks of exposure to a combination of contaminants is usually based on a mathematical (e.g., linear or weighted) combination of the risk of each contaminant separately, but this assumption needs further research.

Risk to the environment refers to, for example, the number of soil microorganisms that do not survive a certain concentration level (i.e., the concentration level at which 25% or 50% of the soil organisms die). Soil organisms are in extensive contact with the contaminants. Hence, at the same concentration level, the risk to soil organisms may be greater than the risk to humans. Especially for metals, risk-related soil standards based on human toxicological considerations are usually less stringent that those based on ecotoxicological considerations (exceptions exist for barium, cadmium, lead, mercury, molybdenum and nickel).

Risk-specific or Function-specific Cleanup Levels

Risk assessments are surrounded with an air of complexity, and indeed, making a detailed risk analysis is complex. Finding and interpreting literature can be difficult. However, for some commonly encountered and well-studied contaminants, using pre-established risk-based action levels and risk-based cleanup levels, on the other hand, make an evaluation of the risk of contamination technically more straight forward. Risk based action levels are established by numerically reversing the steps in an ordinary risk assessment[1]. That is:

1. Assume a target receptor, a known MTR level and behavior of the receptor.

2. Assume one or more exposure routes (paths).

3. Determine the concentrations in the soil that, via the chosen paths and taking the behavior of the target into account, lead to the conditions which do not exceed the MTR. The result of a backwards risk assessment is a list of concentrations, an example of which is given in Table 1[2]. This list specifies increasingly less sensitive functions of the soil, i.e., for children's playgrounds (most stringent), residential areas, park and recreational facilities and commercial and industrial areas (increasingly least stringent).

---

[1] A backward risk analysis may lead to "impossible" concentrations, e.g. when solubility limits are exceeded at a certain risk. The inconsistencies can be eliminated by an iteration process or by artificially limiting upper limit concentrations.

[2] The risk for the environment is defined by statistically manipulating concentration-effect data.

TABLE 1: Function Specific Soil Test Values (Data in mg/kg dry soil weight)

Contaminant (group)	Children's Playgrounds	Residential Areas	Park and Recreational Facilities	Commercial and Industrial Areas
Metals/Metal Oxides				
Arsenic	20	40	100	200
Lead	200	400	1,000	2,000
Cadmium	6	12	30	60
Copper... .	300	600	1,500	3,000
Nickel	60	120	300	600
Mercury	4	8	20	40
Selenium	40	80	200	400
Thallium	0.5	1	2.5	5
Hydrocarbons				
Benzo(a)pyrene	1	2	5	10
Halogen Hydrocarbons				
Hexachlorocyclohexane (mix)	0.2	0.4	1	2
1,2,4-trichloro benzene	5	10	25	50
hexachlorobenzene	0.3	0.6	1.5	3
DDT	0.4	0.8	2	4
PCB	6	12	30	60
Dioxins/furans (ng TE/kg)	60	120	300	600
Phenols				
Phenol	20	40	100	200
Cresols	30	60	150	300
Monochlorophenol	3	6	15	30
2,4-dichlorophenol	2	1	10	20
2,4,5-trichlorophenol	30	60	150	300
2,4,6-trichlorophenol	2	4	10	20
tetra chlorophenol	15	30	75	150
pentachlorophenol	3	6	15	30
Nitroaromatics				
Nitrobenzene	0.6	1.2	3	6
2,4-dinitrotoluene	1.2	2.4	6	12
2,6-dinitrotoluene	0.4	0.8	2	4
2,4-dinitrophenol	1	2	5	10

Source: Hygiene Institute of the Ruhr Area (Foreign Broadcast Service, JPRS Report, Environmental Issues, 15 June, 1994)

Another example is the well-known Dutch list. This list relates to the requirements for a multifunctional soil, which may limit its application to military sites. However, it is not difficult to repeat the exercise for exposure/target scenarios that are specific to a particular function (the necessary information to do this can be found in the various documents that accompany the Dutch list; several simple risk assessment programs are also available).

It is clear that soil that is being used for military purposes can be contaminated at a multiple of the levels required for playgrounds (e.g., the Dutch list C levels) before it poses an actual health risk to military populations which use the site. For this reason, it may not be cost-effective to use multifunctionality criteria for pieces of land that have been assigned specific functions, such as industrial sites and military sites.

Resource-specific Cleanup Levels

The level of contamination allowed in a groundwater extraction area often is based on a requirement concerning the leachability of certain compounds. It can be based on the total leachability of a compound under specified conditions, based on the equilibrium concentration of a toxic substance in water (distribution), or both. When the equilibrium concentrations are less than a specified level (e.g., drinking water standard), the contamination may not be considered threatening, irrespective of its concentration in the soil. Health-related soil concentration levels also may apply to groundwater protection insofar as the concentrations in the two media can be related to each other.

When the direct access to the site is limited or precluded, therefore, multiples of the concentrations in the soil that would lead to direct risk may be allowed, provided that the leachability is low. Although a site with high soil concentrations, but low leachability, would probably be classified as "to be remediated", its remediation would not receive a high priority. The most straightforward technique for the protection of resources is an isolation method (usually hydraulic). Physical barriers can often provide adequate protection, even in the long term.

Action Levels (Contaminant Levels on the Dutch List)

When an action level is exceeded, by definition, an action may be appropriate (but priority of the action may still need to be set). Action levels for remediation can aim at protecting human health, protecting the environment, protecting resources or combinations thereof. Most action levels are human health-related, because, as mentioned earlier, they address the most important criterion. Since the information from preliminary site assessments may show significant variation, action levels which trigger further investigations may be more restrictive than the ultimate cleanup targets. Approaching a risk-based level triggers further investigations to determine whether site-specific risk-based action levels are indeed exceeded.

May 9, 1994, the new Dutch list came into effect [Anonymous, 1994]. The new Dutch guidelines specify only two[3] levels of soil contamination, the A-level (now called the target level) and the C-level (now called the intervention level), respectively. These levels are used to determine the necessity of further investigation and remediation and are, thus, action levels. Function-specific contaminant levels also have been given attention in the Netherlands since they are used to prioritize cleanup projects through risk assessment calculations. However, they are not found in the site assessment protocols since they are not used as trigger values for site assessment or remediation.

C-level The C-level (intervention level) defines the level at which the soil poses a potentially serious danger for public health and the environment. This means that a soil contaminated at the C-level poses a potential health threat for sensitive targets or sensitive organisms when all exposure routes are operational. In other words, to be at actual risk when the soil contamination is at the C-level requires exposure to a child of 15 kg who crawls around in that particular soil. In addition, the contamination must be present in the top 50 cm or so of the soil. The C-level is an action level in that it triggers the need for remediation, though not the priority in which remediation will occur.

A-level The A-level (target level) is a reference value which defines "clean" soil. This level originates from an assessment of a large number of supposedly uncontaminated Dutch soils. Soils having contaminant concentrations below the A-level are considered unrestricted and multifunctional, even though there might have been some contamination. Numerically, the A-level is the 90-percentile concentration (90% of the Dutch soils have relevant concentrations less than the A-value) of the assessed soils. Note that 10% of the Dutch soils are not multifunctional according to this criterion, even though they are considered unpolluted. As a consequence of the method by which the A-level was established, it is country-specific or region-specific. The A-level triggers a suspicion of contamination, but may not lead to remediation.

Action levels typically are established for a range of conditions that are bracketed by site-specific or function-specific health risk levels (maxima) and clean soil levels or analytical

---

[3] The revised guidelines of 1988 listed three levels of pollution (A, B and C) of which only the A-value and the C-value had specific meanings. The B-value was used as a trigger value. In the 1993 revision, the B level has been omitted. It is advised to take a level of B = 0.5*(A+C) as trigger value for further investigation until a better criteriun is available.

detection limits (minima). Remediation is not necessary, by definition, if the health risk level is not exceeded. It is not surprising, therefore, that the discussion about action levels continues. In the Netherlands, the principal action level is the C-level. Once this level is exceeded, the soil has to be remediated, not to the C-level, but to the Dutch A-level. Not all countries have implemented such an approach.

When health-related action levels do not apply, i.e., when the contamination does not pose a threat to human health, use of health-related criteria typically is not relevant (although redevelopment desires may lead to similar criteria). Resource protection requires specific action levels in the same way as does health protection.

Structured Site Assessment

A standard site assessment is an essential element in controlling costs and reaching an acceptable level of cost effectiveness in the remediation program, both at the front end (definition) and at the back end (remediation). Since a decision for further action typically leads to large expenditures, site assessments often are conducted in phases. At the end of each phase, sufficient information must be generated (and not more) to make management decisions concerning the necessity of further actions, the priority of further actions and the planning of funds.

The integrated site assessment/decision system presented in this section follows the general approach used in the Netherlands[4]. This system is not tailored to military sites, and the principles apply to any kind of site. The site assessment system distinguishes three major phases in the investigation, each of which may be subdivided in sub phases as needed. These phases are: Initial Investigation (Phase I), Further Investigation (Phase II), and Remedial Investigation (Phase III) (see Figure 1). Each of these phases is followed by (1) a decision for further action and (2) an optimization of this action. The site assessments influence the decision-making process (and vice versa), the process of resource allocation and the continuity of the project.

---

[4] The site assessments have been protocolized by The Netherlands Organization for Applied Scientific Research (TNO). The decision scheme and the development of action levels were developed by Government Institute for Public Health and Environmental Hygiene (RIVM) and the Ministry of the Housing, Physical Planning and the Environment (VROM).

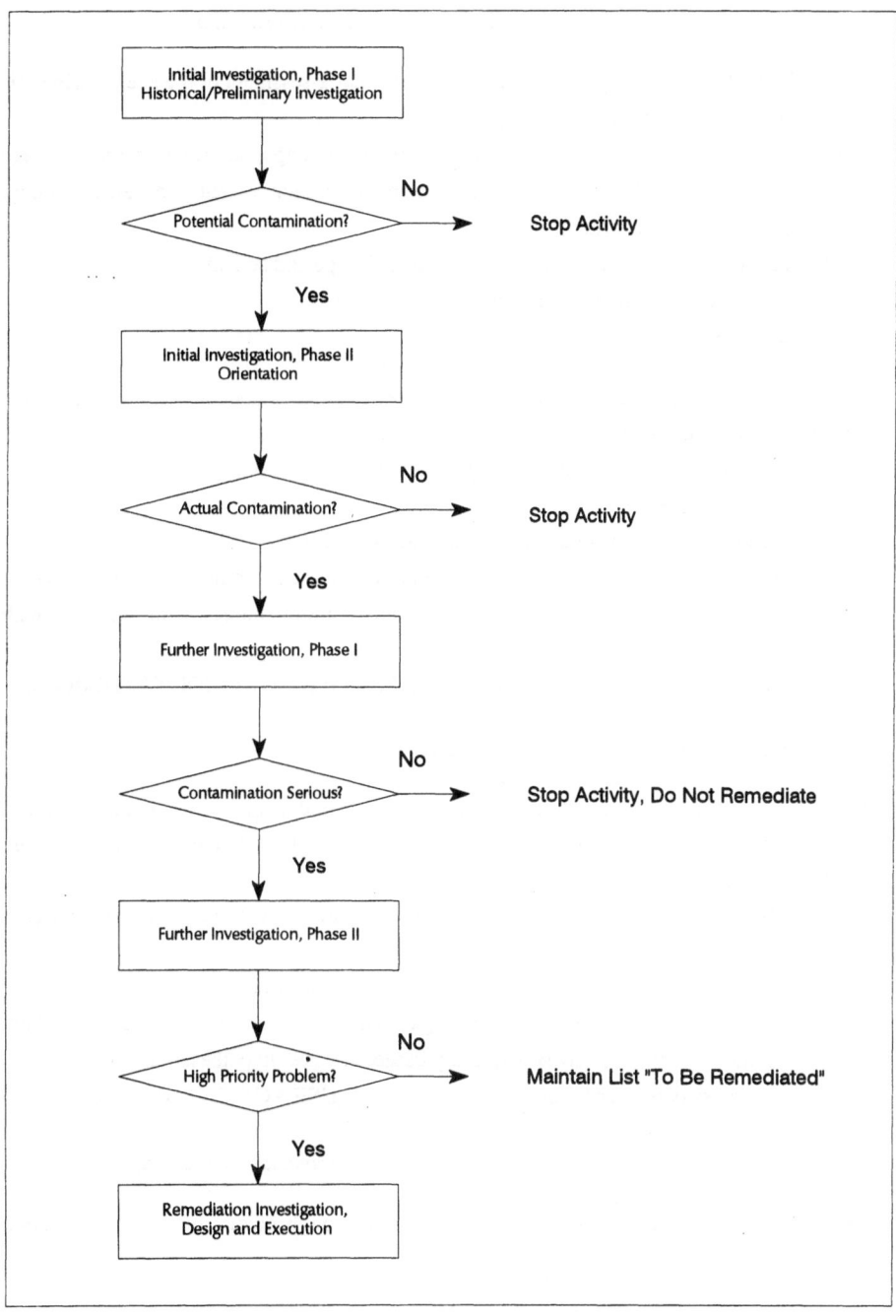

Figure 1: Standard Investigation Scheme with Decision Structure

Sub 1a: Initial Investigation, Phase I (preliminary/historical investigation)

The initial investigation starts with an extensive historical assessment of the site. During this phase, information is gathered concerning:
- the former and present use of the site (old maps, archives, bomb damage assessments, old CIA documents (many of which recently became publicly available) and satellite images;
- the geological and hydrological situation, soil type and soil stratification;
- the surface conditions (visual inspection); and
- the activities surrounding the site.

A visit to the site is part of these investigations (an assessment is not complete without a visit to the site). The historical investigation often leads to statements concerning:
- the contaminants (including unexploded ordnance) that may be expected;
- the size of the contaminated area;
- the location and magnitude of contaminant sources;
- an hypothesis concerning the spread of the contaminants (homogeneous, heterogeneous with known location of one or more sources, heterogeneous with unknown location of one or more sources); and
- the necessity for the Initial Investigation, Phase II (USE ACTION CRITERION).

Sub 1b: Initial Investigation, Phase II (Orientation)

During this phase, the contamination hypothesis developed during the preliminary/historical investigation is used to design a sampling strategy. The results from the sampling round are used to:
- determine whether contamination is actually present (for each location and each source);
- determine if there are additional sources of contamination;
- determine if the hypothesis concerning the contamination spread was correct (if not additional sampling may be necessary based on a new hypothesis); and
- the necessity for Further Investigation, Phase I (USE ACTION CRITERION).

Sub 2a: Further Investigation, Phase I (Determination of Remediation Necessity)

The Further Investigation, Phase I provides a more thorough picture of the contamination situation and focuses on:

- the concentration, total amount and spacial distribution of the contaminants. The sampling strategy is described in protocols [Lamé and Bosman 1992a and 1992b] and is based on hypotheses concerning the local spread of the contamination;
- determination of relevant contaminants that were not included in initial investigation;
- the statistical manipulation and display of sampling data;
- the mapping of sub surface stratification and the geological and hydrological situation; and
- an attempt to relate the present contamination situation to the contaminant sources.

The Further Investigation, Phase I leads to:
- detailed information about the contamination situation; and
- information to determine if there is a need for remediation (USE ACTION CRITERION).

Sub 2b: Further Investigation, Phase II (Determination of Remediation Priority)

In this stage the information generated during the previous stage is used to:
- predict the spread and mobility of the contamination;
- estimate the health risks to humans and the environment;
- estimate the influence of the contamination on the asset value of the site; and
- estimate the possible damage to groundwater resources.

The information of the Further Investigation, Phase II is used to determine:
- the remediation urgency (USE PRIORITIZATION SCHEME); and
- the need for temporary measures, such as hydraulic isolation (USE ACTION CRITERION).

Sub 3: Remedial Investigation

This phase is conducted only when funds for remediation are allocated. During this phase additional sampling may be necessary. The purpose is to provide data sufficient to permit the evaluation of alternative engineering options and, thus, the type of investigation may vary with the available technology.

Decision-making and Action Criteria

For their control, contaminated site projects (projects that include site investigations, remedial investigations and the actual remediation) require a number of "flags," or stop or go points in the project during which the results are evaluated in terms of the project goals or in terms of the entire program. Flags are surrounded with criteria for decision-making, often in the form of action levels, but criteria for cost-effectiveness or program evaluation criteria may also apply. Decision flags are set at those points in the site investigations where the information allows for decision making. The following questions are usually asked:

1. Is there a potential for contamination (suspicion)?
2. Is there in fact contamination (criteria)?
3. Is the contamination sufficiently serious to continue the investigation (criteria)?
4. Should the site be remediated (criteria)?
5. What is the priority of further investigation/remediation (prioritization scheme)?
6. How should the remediation be conducted (list of alternatives)?
7. Has the remediation succeeded (evaluation criteria)?

Event:	Reported suspicion
Issue:	Existing potential for contamination (suspicion)?
Criteria:	There are no formal criteria for a suspicion. The fact that activities known to release pollutants have occurred on a site typically is sufficient. In exceptional cases, a preliminary investigation may lead to the statement that no contamination is believed to be present. In that case the results are reported and no further action is taken.
Decision:	Start the historical/preliminary investigation (Phase Ia Initial Investigation)?

Event:	Completion of historical/preliminary investigation
Issue:	Is there in fact contamination?
Criteria:	In the Netherlands factual evidence of contamination exists when the Dutch A-level is exceeded. Values below the Dutch A-level indicate "non-contaminated" and no further actions are taken. Although this criterion may vary by country, it is important to establish it, since it decides whether or not the site will be placed on the list of "contaminated sites". This same criterion also may trigger the further investigation.
Decision:	Add to list of contaminated sites? Continue with Phase II of the Initial Investigation

Event:	Completion of both phases of the initial investigation
Issue:	Is the contamination potentially serious?
Criteria:	In the Netherlands, the presence of potentially serious contamination, and the necessity of conducting Phase I of the Further Investigation is based on the Dutch "B"-level (which in the new Dutch list is operationally defined as $(A+C)/2$, since a separate B-level is not longer specified). This criterion is rather arbitrary and other criteria can be used instead, such as another concentration limit (e.g., the C-level), the leachability of the contaminants (resource protection), the total mass of contamination (cost optimization per kg contaminant), the potential damage, etc. and various combinations thereof.
Decision:	Add to list of potentially serious contamination? Conduct a Further Investigation, Phase II

Event:	Completion of the Further Investigation, Phase I
Issue:	Should the site be remediated (is the contamination serious)?
Criteria:	The Dutch use a volume concentration criterion to decide whether or not the site should be remediated and if a Further Investigation Phase II should be conducted. This criterion is when 25 m^3 of soil or 40 m^3 exceed the C-level (one sample exceeding C-level does not trigger the need for remediation). If the need for remediation is established, the Further Investigation Phase II should be conducted to establish further information on the contamination situation. An alternative is to use function-specific criteria, e.g., remediation should be conducted when the contamination exceeds function-specific risk-based levels.
Decision:	Add to list "to be remediated"? Continue with the Further Investigation, Phase II?

Event:	Completion of the Further Investigation, Phase II
Issue:	What is the priority of the remediation project?
Criteria:	The prioritization can be based on a health criterion (as it is in the Netherlands), but also on the damage that the contamination may cause to groundwater resources or in terms of the loss of a function of the site. On industrial sites, the use of the site will strongly influence any health-related criteria.
Decision:	Continue with remediation?

Event:	Not specific, depends on priority and available funds
Issue:	How should the remediation be conducted?
Criteria:	This can be a matter of cost-effectiveness, but also one of general philosophy or control. Target-related measures, isolation and cleanup have their limitations and benefits (e.g., hydraulic isolation requires continuous administration and control, whereas cleanup does not). Cost effectiveness will play a large role in this decision. Formulation of formal criteria is difficult. Specific technical questions are answered after the remediation investigation is conducted. Costs of the project are known after a first remediation design is made.
Decision:	Selection of principal remediation strategy.

Event:	Completion of the remediation
Issue:	Has the remediation succeeded?
Criteria:	Evaluation should take place after each project or group of projects. Important evaluation criteria relate to the goals of the program in terms of health protection, groundwater resource protection, site functionality and cost-effectiveness.
Decision:	Continue as before? Choose another contractor? Change the project approach? Change the general approach of the program.

Ranking of Starting Points and Prioritization

The starting points mentioned earlier were non-operational principles that form the basis of a strategy. A strategy is an attempt to realize the resulting goals in the most effective way. This requires making the starting points operational and building a decision structure around them. Not all starting points have equal importance and, therefore, they can be ranked. This ranking forms the basis of an operational prioritization scheme in the remediation strategy.

Priority of Risk Reduction. In an absolute sense, health-related projects receive highest priority, i.e., this criterion overrules all others. The relative priority within a pool of health-related projects involves political and ethical decisions (e.g., is the death of 50 people worse than 100 becoming seriously ill?), which make relative prioritization difficult. Prioritization on the basis of a combination of risk and the number of people affected can be done, although this approach is often quite unsatisfactory.

Priority of Resource Conservation. Drinking water is the single most essential resource for the human species, which makes a drinking water resource a precious commodity. The absolute priority of resource-related problems, therefore, is not much less than that of direct health-related problems. The relative priority of the remedial measures can be derived from the damage the contaminants are expected to cause when not remediated (e.g., cost of installing new drinking water wells, number of people affected). In cases where the drinking water extraction can not be controlled, the criterion of risk reduction prevails.

The requirements for resource protection can be organized in many ways, but all aim at reducing the load of contaminants that moves toward the resource area (e.g., the drinking water extraction area). In general, requirements are formulated in terms of toxicity, leachability (how mobile is the contaminant), the velocity of the contamination front, total mass leachable (how much can be leached), or the mass of contaminant leaving the site. In all cases, the velocity of the pollutants - which is a function of the velocity of the groundwater, the aquifer characteristics and the hydrophobicity of the pollutants - together with the expected damage is a preliminary way of assessing the relative priority of a remediation project.

Priority of Restoring Functionality. The functional use and the asset value of a site can be evaluated using economic analysis. The value of a site depends on the extent to which the site is attractive to developers. Restoring a minimum soil quality may be needed before redevelopment can proceed. To redevelopers, this cost is simply the amount that has to be subtracted from the original (uncontaminated) asset value. Damage and asset value assessments can be used to prioritize sites that have an asset value (are considered for redevelopment). A basis of damage comparison could be the cost to remediate the site to a certain level, e.g., background level (e.g., Dutch A-level), Dutch C-level, or a function-specific level.

From a health point of view, the tolerable contamination of the soil is related to its function. If present land use results in unacceptably high exposure, the priority will be dealt with as an imminent health issue (see previous discussion regarding Sub 1). If the site is no longer used for a specific purpose, the restoration may be a matter of economic evaluation. Since it may be quite costly to restore the soil to admit more sensitive functions (ultimately multifunctionality), changes in the permitted function of a specific site should considered carefully.

Priority of Cost Control. The point of cost-effectiveness is to seek an acceptable solution at a minimum cost not only within a project, but also within an entire program. The criterion of cost-effectiveness, therefore, can result in a re-ranking of projects as it takes the present as well as the future value of each remediation effort into account. This important consideration balances the immediate and future benefits of a remediation effort.

A project that represents a cost-effective solution for one site, may not be cost-effective overall when considering the fact that the need for funds for this project leads to a delay in the remediation of another site (i.e., when the overall budget constraint is considered along with the individual site remediation cost). It often is more cost-effective in the long term to remediate sites that contain mobile contaminants in comparison to sites where the contamination is immobile, especially when the magnitude of the contamination is large.

Cost-improvements can often be attained by considering an approach involving containment (i.e., containment now, cleanup later). Since containment is often less costly than cleanup, you may, for example, be able to implement 3 containment projects for the same cost as one full remediation project. If the benefit is the same (i.e., if both approaches result in the same overall benefit), the containment approach is cost-effective. Moreover, if instead of 3 containment projects, 1 full remediation project was chosen, the other 2 sites would continue to cause incremental problems, perhaps with corresponding increases in overall future costs of the overall remediation.

An evaluation of cost-effectiveness also ensures that projects that do not violate any of the previous starting points, but at some time in the future may require remediation, may in fact be remediated early. A layer of free phase oil on the groundwater in a rural area, for example, even if it does not directly threaten a population or resource, will often be remediated because its remediation in the future would be exorbitantly expensive once constituents become dissolved in water. On the other hand, some highly contaminated sites may actually be cheaper to remediate in the future as new technologies come on line or existing technologies are optimized.

Prioritization of Projects

A ranking process is concerned with several factors: (1) the absolute priority of the defined problem, i.e., its importance in terms of the starting points, (2) the absolute priority of the remediation projects that solve part or all of the defined problem, (3) the relative internal priority of projects that have equal absolute priority, and (4) the maximization of cost-effectiveness on a program management level.

Problems are usually prioritized first, since it is the problems that need to be solved. Absolute first priorities are given directly by the ranking of the starting points. That is, the more important the starting point, the more important it is to solve the problem. Projects can then be prioritized in the order in which they solve high priority problems and/or reduce a high priority problem to a lesser problem by partial remediation. The latter leads in effect to the definition of a new project with a lower priority.

The definition and prioritization of remediation projects is complicated by the dynamic character of the contamination and the sometimes changing usage patterns of the site. Contaminants can be transported with the groundwater, thereby greatly and quickly increasing the area of contamination and hence the cost of remediation. The contamination can threaten to deteriorate groundwater aquifers, and eventually may affect the water supply of distant communities. The mobility of the contamination in the surface and subsurface, the increase in the affected area and the changing functions of the site, may change the relative priority of individual remediation projects.

A relative ranking of projects is based on the future costs and the benefits of the various projects relative to one another. A project can have a higher priority because it is much more cost-effective though it solves a lesser problem (e.g., solving 10 high priority problems for which there are inexpensive solutions can be much more cost-effective than solving one top priority problem for which there is only an extremely expensive solution).

Program Choices and Program Strategy

The magnitude of the program and the choices that are made within the program inevitably are influenced by the available budget. A comparatively larger budget makes it easier to conduct all baseline studies first and subsequently to prioritize the projects. With comparatively limited funds, conducting baseline studies on all sites would absorb the yearly budget for a large number of years without leading to remediations. Politically, this may be difficult to reconcile. If during one year it is decided to only fund initial investigations, the next year it may not be possible to fund all required further investigations on the same sites.

Conducting investigations that will not be followed immediately by further investigations (if they are necessary) is justifiable only if the investigation results are used for prioritization purposes. The following section illustrates some of the basic choices that may be required in the allocation of funds.

Starting points

Simpelland, an imaginary country, has 6,000 suspected sites, of which an estimated 1,500 actually will require remediation. Each suspected site project is conducted in three phases (1) the initial investigation, (2) the further investigation and (3) the remediation, including project design. Remediation can not be completed without being preceded by the first two phases, but not all investigations lead to a remediation. In fact, in Simpelland, 50% of the initial investigations reveal the need for a further investigation, while 50% of the further investigations reveal the need for remediation. Each phase has fixed costs (funds for program execution and management are omitted), see Table 2.

Table 2: Assumed Cost of Investigation and Remediation

Investigation costs	initial investigation	$25,000
	further investigation	$125,000
Remediation costs	target related measures	$50,000
	isolation	$150,000
	cleanup	$450,000

Table 2 shows that the cheapest remediation in Simpelland is the target-related remediation with total cost of $200,000 (total investigation costs and target-related measures). The isolation measure with $300,000 (total investigation costs + isolation), whereas the most expensive remediation is the cleanup with $600,000 (total investigation costs + cleanup). Simpelland has a yearly budget for these remediations of $10,000,000.

Programming Continuous Remediations

The resource manager in Simpelland needs to decide how many projects and of what type he can fund in a given year. Suppose he chooses to initiate a program that will guarantee a continued remediation activity and which funds:

X   initial investigations per year
Y   further investigations per year
N   remediations (N1 target related, N2 isolation, N3 cleanup).

Inserting the cost of initial and further investigation of 25K and 125K, respectively, and the cost of the three options for remediation of 50K (target), 150K (isolation) and 450K (cleanup) we obtain for the total yearly costs:

$$T = 25,000*X + 125,000*Y + 50,000*N_1 + 150,000*N_2 + 450,000*N_3$$

We can combine N1, N2 and N3 by calculating the cost of an "average" remediation, which requires knowing the relative frequency of the various remediation options. Assuming that 20% of the conducted remediations will be a target related remediation (N1 = 0.2N), 40% an isolation (N2 = 0.4N) and 40% a cleanup (N3 = 0.4N), the average remediation cost equals $250,000 (0.2*50,000 + 0.4*150,000 + 0.4*450,000). The total yearly cost then becomes:

$$T = 25,000*X + 125,000*Y + 250,000*N$$

Since it was assumed that 50% of the initial investigations will be followed by a further investigation (X = 2Y) and that 50% of the further investigations will lead to a remediation (Y = 2N), it follows that X = 4N and Y = 2N. Assuming that the currency value is time independent (no interest) and that the available budget for the contaminated sites program remains the same throughout the years, i.e., T = $10,000,000 per year, we can finally write:

$$T = 25,000*4N + 125,000*2N + 250,000*N = 600,000N \leq 10*10^6$$

from which is found N = 16.7. The given budget, hence, is sufficient for the program to fund every year about:

- 16 new remediations of which about 4 are target-related (20%), 6 are isolations (40%) and 6 are cleanups (40%);
- 32 new further investigations; and
- 64 new initial investigations.

Conducting more than 46 further investigations per year will increase the pool of projects labeled as "to be remediated" but decreases the number of projects that actually can be remediated that same year. The projects may take longer than one year to complete.

The Balance Between Investigation and Remediation

If investigation costs are included for sites that were determined not to require remediation, the total cost per actual remediation is $600,000+ (1 remediation, 2 further investigations and 4 initial investigations). Simpelland thus faces a total remediation cost approaching one billion (1,500*600,000 = $900,000,000). If the available budget for remediation is fixed at $10,000,000 (ten million) per year, remediation of all sites takes about 100 years (900,000,000/10,000,000 = 90 years) and the problem would be solved at a rate of 16 remediations per year, assuming no new sites identified.

An alternative is to conduct all initial investigations before starting any remediation. When all funds are allocated to initial investigations (at $25,000 each, see Table 2), 400 of these projects can be conducted per year (10,000,000/25,000). The 6,000 initial investigations to be conducted would require 15 years (6,000/400) to complete and $150 million would be spent on initial investigations, without a single remediation being done. This, alternative is likely to engender little political support.

Comparing both situations makes clear that remediating 16 arbitrarily chosen sites in one year, taking up the budget of a full year, eliminates the initial investigation of 400 other sites for the same year. These investigations could have revealed a number of sites that are rapidly expanding and, if investigated, would have received a higher priority than the 16 remediated. On the other hand, if the 15 years of initial investigations is chosen, some of the worst sites may have caused great problems.

The dilemma of investigating versus remediating sites is not easily solved, but the example illustrates the importance of carefully balancing the number of investigations on the one hand and remediations on the other. A reasonable alternative is to start with the obvious cases first, accepting the consequence that some initially unrecognized problems may become larger by the time they are discovered at some time in the future. The optimum division of work between conducting studies and executing remediation projects is a matter of getting a "feeling" for the situation and should be based on the interpretation of all information available. Getting the most information out of relatively inexpensive investigations is an art well worth practicing.

The increase in cost per year as a result of the expansion of the contamination can be incorporated in the calculations. Such a calculation exercise is a useful tool in the optimization of the program and the evaluation of alternative actions (cost-effectiveness).

References

Anonymous (1994) Beleidsnotitie Interventiewaarden Bodemsanering (policy note intervention levels soil remediation), Tweede Kamer Vergaderjaar 93-94; 22727, nr. 5; SDU publisher, Den Haag.

Berg, R. van den, J.M. Roels (1991) Beoordeling van risico's voor mens en milieu bij de blootstelling aan bodemverontreiniging: Integratie van deelaspecten, Rijksinstituut voor volksgezondheid en milieuhygiene, reportnr. 725201007, Bilthoven, The Netherlands (in Dutch).

Hinsenveld, M., J.W. Assink (1988) Land management at industrial sites, in K. Wolf, W.J. Van den Brink, F.J. Colon (eds.), Contaminated Soil '88, Kluwer Academic Publishers, pp. 505-513.

TCB (1992) Advies herziening leidraad bodembescherming I: C-toetsingswaarden en urgentiebeoordeling, Technische Commissie Bodembescherming, Leidschendam, The Netherlands (in Dutch).

TCB A01 (1992) Advies Herziening leidraad bodemsanering I: C-toetsingswaarden en urgentiebeoordeling, Technische Commissie Bodembescherming, Leidschendam, The Netherlands (in Dutch).

Lamé, F.P.J., R. Bosman (1992a) Leidraad bodembescherming: Orienternd onderzoek naar di aard, concentraitie en omvang van bodemverontreiniging, TNO report R/92/112.

Lamé, F.P.J., R Bosman (1992b) Leidraad bodembescherming: Nader onderzoek naar de aard, concentraitie en plaats van vookomen van bodemverontreiniging, TNO report R92/113.

INNOVATIVE CHARACTERIZATION TECHNOLOGIES TO

ADDRESS ENVIRONMENTAL PROBLEMS AT U.S. DEPARTMENT

OF ENERGY SITES

Caroline Purdy
U.S. Department of Energy's Office of Technology Development,
Office of Environmental Restoration and Waste Management
1000 Independence Ave.
Washington D.C., 20585
USA

David Roelant
BDM Federal Inc.
555 Quince Orchard Rd
Gaithersburg, MD 20878
USA

The U.S. Department of Energy (DOE) is the governing agency of the Nuclear Weapons Complex, an industrial network of national laboratories and production facilities created to conduct research and development, and to test and produce nuclear weapons. Throughout 45 years of production, the Complex has assembled tens of thousands of warheads and generated millions of cubic meters of hazardous and radioactive waste. Some hazardous contaminants have been dumped, leaked, buried or injected into the ground and exist in concentrations well above drinking water standards at several sites. In addition, aboveground and underground weapons testing has contaminated large areas of land and hundreds of properties involved with the original Manhattan Project and thousands of other facilities in need of decontamination. Finally, thousands of sites that utilized uranium mill tailings for fill material also need to be remediated.

Characterization technology is critical at contaminated sites for assessing the extent of contamination, quantifying the risk to the public and to site workers, selecting the appropriate remediation plan, monitoring the remediation, and providing required site closure monitoring.

NATO ASI Series, Partnership Sub-Series, 2. Environment – Vol. 1
Clean-up of Former Soviet Military Installations
Edited by R. C. Herndon et al.
© Springer-Verlag Berlin Heidelberg 1995

For decontamination and decommissioning (D&D), characterization technologies are needed to: assess the physical safety of facilities (many long since abandoned), measure the general extent of contamination to provide a basis for the D&D plan (e.g., tear down, entomb, clean sufficiently for future waste operations, clean for transition out of Weapons Complex), and monitor D&D activities to assure worker safety and the effectiveness of the decontamination process. The judicious choice of characterization technologies can save billions of dollars in remediation and D&D costs. The costs for characterization technologies range from tens of dollars for immunoassay kits to nearly a million dollars for Cone Penetrometer Trucks (CPT) with contaminant sensors. Since this is a program of technology development, we will focus on sensors currently undergoing research, development, demonstration and testing and not upon the thousands of commercial sensors. Although some sensors are pertinent to Weapons Complex sites only, most are applicable at industrial sites as well as federal facilities. The three areas of technology described in this paper, each with enormous potential cost savings are: field-deployable instruments for site and facility characterization, CPT sensors, and characterization methodologies.

Field-deployable geophysical instruments are useful for rapid identification of subsurface structures to help identify hydrogeologic features, locations of pools of dense non-aqueous phase liquids (DNAPL) and optimal locations for siting remediation wells. The oil and gas industry and the mining industry have often led in the development of these techniques. The spatial and temporal resolution of many environmental applications are more demanding than those developed for resource exploration. Geophysical instruments are robust, reusable, and often require extra data collection and analysis costs but not capital costs. Technologies under development include: three dimensional and three component seismic imaging, cross-well seismic tomography, zero tension lysimeters, synthetic aperture radar, ground-penetrating radar and electrical resistivity tomography. Since several facilities that have ranges for testing and evaluating geophysical instruments are heavily utilized, and since many are geared toward specific U.S. Department of Defense (DoD) applications (e.g., mine detection), DOE has developed a geophysical test range in Rabbit Valley, Colorado. The site is easily accessible from the highway and contains buried objects of various shapes, compositions, sizes and spacings designed to address resolution requirements of environmental applications.

Field-deployable chemical instruments are needed to rapidly assay contaminant concentrations in soil, waste, groundwater, air and on facility walls, floors, equipment, pipes, etc. The instrument types are highly variable and are driven by the various contaminants and characterization needs. More data can be analyzed in the field for much less money than when using detailed laboratory analyses. Higher sample throughput, faster extraction procedures and streamlined analysis methods all contribute to same day results (versus weeks to months for lab results). For elemental (including radionuclides) assays: Inductively-Coupled Plasma (ICP) Mass Spectroscopy (MS), ICP - Atomic Emission Spectroscopy, Laser-Induced Breakdown Spectroscopy, borehole multispectral neutron logging, a high-energy beta particle sensor, and an electrometer for collecting electrically charged air molecules as a gross measurement of alpha-emitting radionuclides are under development. Field assays of organic compounds can be conducted by Ion Trap MS, Ion Trap/Gas Chromatography/MS, acoustic wave chemical sensors, Secondary Ion MS, and Laser-induced Fluorescence systems.

The Cone Penetrometer Truck is a heavy vehicle that is capable of hydraulically pushing a hollow steel rod (i.e., cone penetrometer) into the earth to depths exceeding 200 feet. It was originally developed in Europe decades ago. The original CPT measured the sleeve and tip friction at the end of the cone penetrometer and was able to classify the thickness and soil type of the various geologic layers. In the past decade, the U.S. Army Core of Engineers, DOE national laboratories and a few U.S. vendors for the CPT began developing sampling devices to retrieve and analyze soil, groundwater and soil gas for the presence of contaminants for environmental applications. In the past few years, the technological advances for the CPT have dramatically increased, as remediation projects at several DoD, DOE and U.S. Environmental Protection Agency (EPA) contaminated sites have been used to demonstrate its use. For example, it is standard practice at contaminated sites for the CPT to grout as the cone penetrometer is withdrawn, thereby preventing new migration pathways for the contaminants.

The development of sensors and sampling devices engineered into the end of the cone penetrometer is currently an area of significant activity in the United States. Funding agencies for these developments include: DoD, DOE, and the Advanced Research Projects Agency (ARPA). Sensors under development by DOE include: time-domain reflectometers, fiber optic chemical sensors, scintillation and solid state radiation detectors, a positioning device for exact penetrometer tip location, laser-induced fluorescence instruments for organics, moisture detectors for the vadose zone, fiber optic sensor for pore pressure, and a chemiluminescent detector for decomposition products of explosives.

In addition to the development of sensors for the CPT, screening and quantitative field analytical methods coupled to the CPT are under development, such as methods for: on-line, near-real-time measurement of volatile organic compounds (VOCs) and semi-VOCs, GC/MS instruments, preconcentration devices to improve sensitivity, thermal desorption for sample introduction, heated transfer lines for minimal adsorption of contaminants during sample transport, and inert carrier gases for transport and entrainment of contaminants. Finally, a review of sensors and instruments available for the CPT was performed. The published report describes the history of the CPT, typical subsystems, sensors and instruments available, sensors and instruments under development, operators of CPTs in the U.S., review of books and papers on CPT applications and a list of references.

The CPT can be used in many soil types including: sand, clay and small-grained gravel. By adding extra weight to the CPT (40 to 60 tons) the pushing capability of the CPT was improved from 6 feet to over 100 feet in the large unconsolidated cobbles at the DOE Hanford site. A sonic drilling head incorporated on the penetrometer is under development and is expected to further extend the CPT's versatility. Although a CPT can cost anywhere from $0.4-1.2 million (USD), its potential cost savings can be enormous as evidenced by the cost analysis discussed in what follows. By providing a detailed subsurface characterization of a contaminated site, the CPT can help eliminate poor remedial design and the inappropriate placement of monitoring and remediation wells. In the U.S. the labor cost for running a CPT is approximately 14% of the total cost. Eventually much of the labor cost and worker exposure could be reduced through the use of robotics. In other countries the labor cost is much lower and robotics may not be an economically feasible alternative at this time. The cost of using the CPT to complement drilling compared to drilling alone is analyzed below for five different hypothetical sites. We look only at the cost savings from avoiding drilling and installing extra monitoring wells. The savings from the optimal placement of remediation wells and the overall improved remedial design possible with better subsurface characterization of contaminant plumes are hard to quantify and not considered in this cost savings analysis.

At site A the drilling method is mud rotary, depths are 75 feet, and the contaminant is trichloroethylene. The conventional site characterization plan calls for 20 monitoring wells. Using the CPT to sample at 40 locations, the site characterization plan requires only 10 wells. The cost for drilling and well installation for the conventional plan is $278,000 (USD) versus $143,000 (USD) for the plan using the CPT. This represents a cost savings of 49%. In addition, the optimal placement of the ten monitoring wells provides much better data than the conventional plan's twenty wells. The "break even" point for this site occurs when 15% of wells are avoided.

At site B a hollow stem auger and a depth of 50 feet are required. Drilling costs per well vary from no core ($7,687 (USD)) to continuous coring ($10,087(USD)). The break even point at this site occurs for 10% of wells avoided, and the cost savings are 55% ($275,700 (USD) versus $124,900 (USD)). Since the time required to set up a drilling rig exceeds that for the CPT, the economics of shallow investigations favor the CPT.

For site C, we consider the preliminary assessment (initial look at a site). When there is no information available for a site, it is common practice to install four wells. The data from sampling soil and groundwater at various depths with the CPT yields more information than that from drilling (even continuous coring). Depending upon site conditions, anywhere from 40 to 90 CPT pushes can be done for the same cost as the four wells ($73,000 (USD) including 1 year monitoring costs).

Site D is a very large area site. The number of wells in the conventional site characterization plan is 100. The number of CPT push locations used is 200. Mud rotary drilling is assumed to a depth of 150 feet. The break even point occurs at 25% of wells avoided. Site E is simply site A with the presence of high levels of radioactive and hazardous contaminants which require extra procedures, training, time and costs for both drilling and using the CPT. For drilling operations there are more workers and they are exposed to much higher levels of contaminants than CPT operations. This results in an overall lower break even point than site A (e.g., less than 15% wells avoided).

In the U.S. where regulations abound, a conservative estimate of wells that can be avoided is 30-50%. From a statistical and scientific viewpoint, much higher than 50% of wells can be avoided. For Eastern European countries strapped for cash and not mired in regulations, the possible cost-savings are staggering.

A close examination of numerous site characterizations throughout the DOE complex and throughout the United States reveals that the cost, the length of time and the scientific proficiency of many site characterizations are hindered less by the availability of new technologies than by the implicit acceptance that site characterizations should be driven mostly by regulations. Defensible science within a regulatory framework should drive the process.

It is our experience that the biggest savings possible in the environmental restoration and waste management field is provided by implementing innovative methods which integrate a judicious choice of advanced technologies into a common sense approach. For site characterization and remediation there is a methodology entitled Streamlined Approach For Environmental Restoration (SAFER) which was developed by the EPA and the DOE. This methodology will not be discussed further, except to mention that The Expedited Site Characterization methodology fits well into the characterization stage and that the methodology examines data quality objectives which help eliminate the collection of useless or minimally useful data. In the Characterization Program, we continually examine the characterization needs at DOE sites and strive to develop characterization methodologies which hold great promise for reducing characterization costs. Characterization methodologies currently under development are:

- a mixed waste treatment plant process control diagnostic monitoring system, to precharacterize waste for treatment selection, monitor organic destruction, monitor all effluents, monitor energy costs for treatment, and monitor the integrity of the final waste form;
- a data fusion computational tool for site characterization which can combine the data from several disparate data sets into a single model that allows an iterative modeling approach (e.g., electromagnetic and core data can greatly increase the precision of the seismic data and yield a vastly improved subsurface characterization);
- an Expedited Facility Characterization (EFC) is still under early development, but seeks to address the physical safety of facilities in addition to the characterization of the hazardous and radioactive contaminants contained therein;
- field analytical methods for chemical instrumentation for rapid, accurate analysis of soil, waste and groundwater samples in the field; and
- the Expedited Site Characterization (ESC) methodology for timely and inexpensive characterization of contaminated sites.

The ESC methodology has been successfully implemented at several sites, including Comprehensive Environmental Response, Compensation and Liability Act (CERCLA) sites, and shows enormous potential for cost- and time-savings. Argonne National Laboratories initially developed the ESC methodology and continues to improve upon it. The methodology is a common sense approach that does not rely upon any specific technology but has a goal of maximizing information while minimizing sampling and overall costs. The ESC methodology integrates the field-deployable instrumentation and the CPT (where possible) with the entire site characterization plan and contains the following elements:

- a critical evaluation and use of existing site data;
- a site visit for logistical purposes prior to field work;
- early and regular meetings with the local public and the regulators;
- a selection of appropriate technologies (minimally-intrusive, field screening analytics), and multiple technologies for data corroboration;
- a dynamic work plan;
- the entire scientific team working together in the field;
- daily data reduction and integration;
- modification of the field program as necessary for optimization; and
- a final report and acceptance by regulators (e.g., signing of a Record of Decision).

The two critical elements of the ESC methodology are the dynamic work plan and the multidisciplinary team of scientists in the field making decisions for sampling locations based upon daily data integration into a computer visualization model.

The work plan is truly dynamic. The only locations for sampling determined prior to arriving at the site are those of the first day's data collection. Thereafter, hydrology, chemistry, daily data (e.g., science) drive all further sampling decisions and activities.

A team of scientists with diverse expertise and strong field experience (geologists, hydrologists, chemists, biologists, geophysicists, geochemists, computer scientists, health and safety staff, regulatory staff and a technical support staff) is essential. The technical team works together throughout the process from initial work plans to final data analysis and reporting. The technical team first critically reviews and interprets all existing data to determine which data sets are technically valid and can be used in designing the initial field program. One of the most repeated mistakes made in site characterizations is the poor utilization of available data for forming initial hypotheses and initial testing. Most important is that the experienced scientists are in the field analyzing data and making decisions concerning sampling on a daily (sometimes hourly) basis.

The cost savings associated with deploying the ESC methodology include not only that for the CPT (avoiding monitoring wells) and the field-deployable instruments (more rapid and lower cost than lab analyses) but also the savings from integrating the entire characterization data collection process. The ESC methodology has been used to characterize U.S. Department of Agriculture (USDA) sites contaminated with carbon tetrachloride and chloroform. A site in Murdock, Nebraska was characterized in months with only 3 weeks of field activities costing $170,000 (USD). A comparable USDA site characterized using the conventional method of drilling wells and sending samples to a laboratory for analysis had field work costing over $1.7 million (USD) spread over years. For large, complicated federal sites, regulators tend to be extremely conservative and federal contracting long and expensive. These sites have a higher overhead cost percentage, tend to be less accepting of innovative technologies and more conservative in assessing the need for additional characterization data. Even under this worst case scenario, ESC is expected to cut costs in half.

In summary, the capital and operating costs associated with characterization technologies vary dramatically. The choice of technologies appropriate for a given site is critical. The cost savings achieved using the best technologies will quickly repay both capital and operating costs. It is important not to forget that there is typically a large cost associated with not characterizing a site properly prior to beginning the remediation activities. Finally, the purchase of new technology is not paramount. The appropriate use of existing innovative technology in the framework of a common sense approach (methodology) to solving problems is of greatest importance.

Caroline Purdy
U.S. Department of Energy's Office of Technology Development,
Office of Environmental Restoration and Waste Management
1000 Independence Ave.
Washington D.C., 20585
(301) 903-7672
fax (301) 903-7457
Dr. Purdy is the program manager for the Characterization, Monitoring and Sensor Technology - Integrated Program and is responsible for establishing and managing a national applied R&D program to develop technologies for site and waste characterization. She received a Ph.D. in Chemistry from the University of Maryland concentrating in Geochemistry and nuclear analytical techniques.

David Roelant
555 Quince Orchard Rd
BDM Federal Inc.
Gaithersburg, MD 20878
(301) 212-6218
fax (301) 212-6251
Dr. Roelant is a senior scientist in the Environmental Sciences Division of BDM Federal. Mr. Roelant has 12 years experience developing a wide range of sensors, theoretical simulation models and imaging processes. He ran a small high technology firm before moving to BDM Federal.

SELECTED APPLICATIONS OF BIOREMEDIATION IN HAZARDOUS

WASTE TREATMENT

Katalin Perei and Béla Polyák
Institute for Biotechnology
Z. Bay Foundation
Derkovits fasor 2
H-6726 Szeged
Hungary

Csaba Bagyinka, Levente Bodrossy and Kornél L. Kovács*
Institute of Biophysics
Biological Research Centre
Hungarian Academy of Sciences
Temesvári krt. 62
H-6726 Szeged
Hungary

Principles

A. Controlled Mixed Cultures

Environmental contamination usually consists of a mixture of dangerous chemicals which, to be degraded, will require a complex ecosystem of microbes. While none of the individual member species of this ecosystem are generally capable of decomposing all of the components of the mixed pollution, the concerted action of various species can potentially bring about the desired cleanup effect.

Effective bioremediation technologies should, therefore, invoke a mixture of microorganisms forming synergistic consortia. A principle dilemma for remediation microbiology is that the base of knowledge on co-operative effects and interactions among microorganisms is rather limited, since most basic research experiments require the use of pure strains. This requirement necessarily excludes any observation of potential beneficial aspect of interactions among various strains.

---

*   To whom correspondence should be addressed

NATO ASI Series, Partnership Sub-Series, 2. Environment – Vol. 1
Clean-up of Former Soviet Military Installations
Edited by R. C. Herndon et al.
© Springer-Verlag Berlin Heidelberg 1995

This lack of fundamental applied environmental microbiological knowledge leads to poorly understood system characteristics and inferior or unpredictable bioremediation performance. Additionally, system control in most applications is restricted to regulation of nutrients. One technique used in several laboratories, including ours, is the assembly of controlled mixed cultures, in which pure cultures of bacteria are deliberately mixed in order to enhance bioconversion/biodegradation yields. To further intensify the microbiological activity, immobilization techniques can be used. This increases the active biomass concentration and may help prevent the escape of cells from the reaction volume as well (Caplan, 1993; Liu and Suflita, 1993).

B. Selection pressure by rare substrates

The general approach to developing bioremediation systems is based on the assumption that naturally occurring microbes exist which are capable of degrading even the most noxious chemicals, albeit at a slow rate. It follows from this principle that microorganisms particularly adapted for decomposing certain types of contamination will be most abundant with respect to the total microbial population *at the site of contamination*. The more incompatible a given chemical is to common forms of microbial life, the stronger the selection pressure is for microorganisms that are capable of metabolizing that compound (Blackburn and Hafker, 1993; Lowe et al., 1993).

Applications

A. Utilization of hydrogen metabolism in biotechnological applications

Hydrogen evolution by intact bacterial cells is frequently observed in nature. In microbial ecosystems, the role of these bacteria is to create and maintain an anaerobic, reducing environment, and to supply the universal reducing agent, molecular hydrogen. Gaseous hydrogen is usually not released from these natural ecosystems unless there is an excess of reducing power which needs to be counteracted in order to ensure optimum metabolism and growth equilibrium in the population. H_2 generated *in vivo* by hydrogen forming bacteria is utilized by hydrogen consuming members of the microbial ecosystem. Hydrogen is transferred very effectively by interspecies hydrogen transfer. The molecular details of this process are not fully understood, but their significance in maintaining both the optimum performance of the entire ecosystem and the delicate regulatory mechanisms should be appreciated. In the mixed population bacterial systems presented here, the advantages of interspecies hydrogen transfer are exploited.

A.1. Biogas production from wastes

Biological methane production is carried out in three stages, each performed by a separate group of microorganisms. Complex organic materials are first hydrolyzed and fermented into fatty acids by facultative and anaerobic microorganisms. Next, the fatty acids are oxidized to produce H_2 (dehydrogenation) and organic acids (acetogenesis), primarily acetate and propionate. The last stage is methanogenesis. Some methanogenic bacteria can combine the hydrogen with carbon dioxide to form methane; others split the acetate into carbon dioxide and methane (Hall et al., 1992).

Among the significant recent advances in understanding the ecology of anaerobic biodegradation of organic wastes is the recognition of the close syntropic relationship among the three distinct microbe populations and the importance of H_2 in process control (Benstead et al., 1990). One can manipulate hydrogen levels and, hence, interspecies hydrogen transfer in order to optimize the concerted action of the entire population. For example, the concentration of either acetate or hydrogen, or both together, can be reduced sufficiently to provide a favorable free-energy change for propionate oxidation.

During anaerobic biodegradation, hydrogen concentration is reduced to a much lower level than that of acetate. The acetate concentration in an anaerobic digester tends to range between 10^{-4} and 10^{-1} M, while H_2 ranges between 10^{-8} and 10^{-5} M, or about four orders of magnitude less. In addition, the hydrogen partial pressure can change rapidly, perhaps varying by an order of magnitude or more within a few minutes. This is related to its rapid turnover rate. The amount of energy available to the acetate-using methanogens is independent of hydrogen partial pressure; the opposite is true for the hydrogen-producing and hydrogen-consuming species.

Low concentrations of free hydrogen within the digester affect the rate of H_2 turnover in the system. In a typical system, the turnover rate of the hydrogen pool is 6.8×10^6 per day, or about 80 per second. The interspecies H_2 transport is therefore a very rapid and delicate rate-limiting step. For microscale processes such as this, diffusion is the main mechanism for hydrogen transport between species. Calculations clearly indicate that the optimal distance between interacting hydrogen producing and consuming species is about 10 microns under practical conditions, or about 10 bacterial widths (Kovacs and Polyak, 1991).

Clearly the bacteria must indeed be close together, otherwise interspecies hydrogen transfer will be the rate limiting step for the overall process. This is the case for the ideal, completely mixed or continuously stirred tank reactor where the individual bacterial species are dispersed uniformly throughout the system, and the distribution of reactants, intermediates, products of biotransformation, and bacterial species is homogeneous throughout the reactor.

New process designs such as the biofilm reactors (Van Loosdrecht and Heijnen, 1993), encourage different species to live in close proximity to one another. The result is a much higher rate of conversion per unit volume of reactor, though this technique has not yet been optimized. Biofilms also permit a diversity of environments on a single biofilm, so that suitable conditions for the oxidation of each substrate and intermediate can exist somewhere within the biofilm.

We have shown that under these circumstances, the addition of hydrogen producers brings about advantageous effects for the entire methanogenic cascade. The decomposition rate of the organic substrate (animal manure in our previous experiments) is increased, and both the acetogenic and methanogenic activities are remarkably amplified. In laboratory experiments, a 2.6-fold increase in biogas production has been routinely observed. The same results were obtained in a 1 cubic meter digester scale-up experiment.

Achieving increases in biogas formation rates from municipal wastes is an ongoing effort. Municipal solid wastes are typically difficult to break down, and biodegradation is particularly slow in the landfill type of "bioreactors." Nevertheless, a cost-effective biogas producing system has been installed at the municipal waste storage field for the city of Szeged. This City Council facility receives about 200,000 m^3 of wastes annually, mostly solid household garbage. The biogas has been collected through an underground pipeline system, compressed, then fed into a natural gas-based heating center that serves a residential area. Biogas production had reached 700,000 m^3 by 1990. The biogas collection and utilization system is simple and inexpensive, and it operates smoothly and efficiently.

Proper management of the bacterial population is expected to facilitate fermentation. In order to reduce the costs of this treatment, supplemental bacteria are grown in diluted industrial wastewater. Spraying the microbes onto the surface of the garbage layer before covering the layer with dirt accelerates the formation of biofilms, the highly effective microbial centers of fermentative degradation.

A.2. Denitrification

Nitrate contamination of natural waters is gradually increasing worldwide especially in the industrialized nations. Several methods have been considered for removing nitrates from water. Currently, ion exchange is the only effective technique used in full-scale treatment facilities. With continuous regeneration of the ion exchange resin, it is possible to clean groundwater containing 20-30 mg per liter of nitrate-nitrogen to lower than 2 mg per liter. Several ion exchange systems are in operation worldwide, with mixed results in terms of system performance and efficiency.

Biological denitrification has been studied for the purification of both wastewater and drinking water. The current technique uses denitrifying microorganisms in a filter bed which reduces the nitrate to nitrogen gas. The biological filter is followed by a conventional filter to remove the excess bacterial growth. The organisms require an organic energy source. Therefore one must be added when treating drinking water, since groundwater supplies are essentially free of organic material. In spite of its success in eliminating nitrate contamination, biological denitrification has not yet been enthusiastically accepted and utilized by water treatment facilities.

The system developed in our laboratories includes two features which, to our knowledge, have not been applied in denitrification technologies before. First, we intend to exploit the benefits of interspecies hydrogen transfer. Second, we hope to use this in conjunction with our novel immobilization technology.

Interspecies hydrogen transfer is essential for supplying the necessary hydrogen for nitrate reduction by denitrifying microorganisms. In our approach we employ well-defined mixtures of bacteria. One species in this mixture is responsible for hydrogen production from added organic substrates, such as industrial wastewaters, sugars, or cellulose. The immobilization technology allows growth in close spatial proximity, leading to efficient hydrogen transfer from the helper microorganisms to the denitrifying bacteria (Kovacs and Polyak, 1991).

The advantages of this system are numerous. Interspecies hydrogen transfer is a very efficient way to administer hydrogen to the denitrifying bacteria. From an operational safety standpoint, *in situ* generation and utilization of hydrogen is superior to bubbling the system with highly explosive hydrogen gas. Immobilization of the participating bacteria in beads of high physical resistance provides higher efficiency due to dramatically increased bacterial population densities and a high flow rate. Moreover, our unique immobilization technique leads to essentially sterile fermentation conditions which allows for strict control of the microbiological ecology within the immobilized system.

In its simplest form, the process contains an ion exchange column producing potable water, while another ion exchange column is regenerated through a biological denitrification reaction. During ion exchange resin regeneration, a brine solution containing nitrate in high concentration is formed. A suitably engineered bacterial population is able to convert nitrate to nitrogen gas. In this way, the regenerant can be recycled after it has been subjected to denitrification, thus minimizing salt requirement and brine production. In laboratory experiments, a 90% reduction in brine volume has been routinely achieved.

Biological regeneration of the brine solution from nitrate-loaded anion-exchange resins can be achieved in the bioreactor containing immobilized microorganisms. In the system developed in our laboratory, a closed recirculation circuit is used, in which the regenerant is recirculated through the ion exchange column (which has to be regenerated), and a detnitrification bioreactor.

In the bioreactor, the bacterial populations are co-immobilized in an alginate matrix, which facilitates both interspecies hydrogen transfer and efficient nitrate reduction. There are a variety of helper bacteria capable of producing hydrogen from simple organic substrates. The best hydrogen producer helper bacterial strains can utilize the organic material found in dilute wastewaters from the food processing industry (sugars and proteins). Thus, treatment of polluting waste waters can be linked to the elimination of nitrate contamination. Bacteria for this system come from a wide variety of common bacterial strains, including the genera *Pseudomonas*, *Achromobacter*, and *Bacillus*. In our hands, such a biological denitrification system can handle 1,000-2,000 mg nitrate per liter (100 mmol $NaNO_3$ per liter) at a conversion rate of 10 mg nitrate per gram of beads every hour. Even after several weeks of continuous operation, nitrite was not detectable in the effluent of our experimental bioreactor (Kovacs and Polyak, 1991).

B. Degradation of hazardous chemicals by natural isolates
B.1. Halogenated hydrocarbons

Methanotrophic microorganisms oxidize methane to carbon dioxide and water in four steps, a process involving methanol, formaldehyde, formate as intermediates. It is possible to separate the pathway into four steps in the cell free extract or after partial purification of the various enzymes. The key enzyme is a metalloenzyme, methane monooxygenase (MMO), which catalyzes the oxidation of methane to methanol.

Methanotrophs have great biotechnological potential in other environmentally crucial bioconversion reactions as well. MMO catalyses the incorporation of oxygen into a broad range of substrates including alkanes, alkenes, aromatic hydrocarbons, alicyclic hydrocarbons, halogenated alkanes, and halogenated aromatic hydrocarbons (Murrell and Dalton, 1992).

Groundwater contamination by chlorinated solvents, such as trichloroethylene (TCE), presents serious health problems. Widespread use and careless handling, storage, and disposal of these chemically stable halogenated hydrocarbons make them among the most frequently detected groundwater contaminants in many industrialized nations. Contamination of drinking and well water cause great concern because halogenated organics can be both toxic and carcinogenic (Ensley, 1991; Bouwer and Zehnder, 1993).

There are several methods available to remediate groundwaters poisoned with organic compounds. Surface waters and groundwater can be treated by adsorption onto charcoal, gaseous-phase air stripping, chemical oxidation or biological processes. However, adsorption, air stripping, and precipitation only transfer the dangerous compounds to other sites in the environment. Biological and chemical destruction processes are advantageous since they lead to the partial or complete destruction of contaminants (Hutchins, 1991; Mohn and Tiedje, 1992; Bowman et al., 1993).

It has been known since 1985 that methanotrophic microorganisms may be involved in aerobic TCE metabolism. Haloalkane oxidation is accomplished though intramolecular hydrogen or hydride migration. This is similar to reactions observed with cytochrome P-450. The iron cluster of MMO generates an activated oxygen species with reactivity similar to the heme of cytochrome P-450. A comparative survey has suggested that any microbial oxygenase can catalyze the oxidation of TCE, though substrate specificity and reaction rates may vary (Murrell and Dalton, 1992).

The data available to date show considerable promise for the eventual biological treatment of these groundwater contaminants. An oxidation rate of 150 nmol per minute for each mg protein (measured with whole cells of *Methylosinus trichosporium* OB3b expressing MMO) has been measured, meaning that a suspension of this bacterium at 1 mg/ml protein would eliminate 20 mg/liter (20 ppm) TCE, a fairly high contamination level, within 1 minute. This rate is very appealing for practical applications. However, this ideal degradation rate contrasts with the persistence of TCE in the environment, emphasizing the difficulty of applying laboratory degradation rates to field conditions. At least two problems have emerged from these studies. First, in order to achieve a steady and fast halogenated hydrocarbon degradation, excess reducing power must be provided in the cells to sustain their redox balance which is disturbed by the oxidation of halogenated compounds. Second, the MMO active site is apparently rendered inactive by the very metabolites generated by the oxidation process itself (Tsien et al., 1989).

B.2. Sulfanilic acid

Sulfanilic acid is representative of aromatic sulfonated amines widely used and manufactured as intermediates in the production of azo dyes, plant protectives and pharmaceuticals. Its degradation is slow and incomplete in most biological systems because the sulfonic acid group is a xenobiotic structural element. In addition, being an extremely strongly charged anion restricts penetration into intact bacteria. As a consequence these compounds are persistent in aerobic wastewater treatment plants. Additionally, sulfanilic acid is the intermediate for the synthesis of various sulfonamid drugs, noted for their strong bactericidal properties. This effect results from their capability to inhibit nucleotide biosynthesis.

Biodegradation of sulfonated aromatic compounds has been studied for about 40 years (Feigel and Knackmuss, 1988, 1991), but the only reactions with published experimental support involve dioxygenases or monooxygenases acting on the carbon atom carrying the sulfonate moiety (Feigel and Knackmuss; 1988, Locher et al., 1991). By employing oxygenases in natural microbiological systems capable of degrading halogenated hydrocarbons, sulfonated aromatics can be utilized through strategies similar to other bacterial activities in the environment, e.g., decomposition of aromatic or aliphatic hydrocarbons. The common element of this strategy includes a first attack on the chemically inert target molecule by enzymatic oxygenation. Oxygenated derivatives are less stable and are more amenable to further chemical and/or biological degradation into products that are easy to metabolize (Locher et al., 1991; Goszczynski et al., 1994).

In our effort to supplement the wastewater treatment system of Nitrokémia Ltd., (a major Hungarian chemical plant producing sulfanilic acid in large quantities), with a biological sulfanilic acid-eliminating reactor, bacterial strains capable of effectively degrading sulfanilic acid have been isolated and immobilized. The optimized system appeared to function effectively at laboratory scale, and industrial experiments are in progress.

Acknowledgments: The experimental work discussed in this paper was supported by several grants from the OTKA, OMFB, PHARE Accord programs and from the U.S.-Hungarian Joint Fund. The authors are thankful for the financial assistance provided by these agencies.

References

Benstead J, Archer DB, Lloyd D (1990) Role of hydrogen in the growth of mutualistic methanogenic cultures. In: Microbiology and Biochemistry of Strict Anaerobes Involved in Interspecies Transfer (Eds. Belaich JP, Bruschi M, Garcia JL) pg. 161.

Blackburn JW, Hafker WR (1993) The impact of biochemistry, bioavailability and bioactivity on the selection of bioremediation techniques. TIBTECH 11:328-333.

Bouwer EJ, Zehnder AJB (1993) Bioremediation of organic compounds - putting microbial metabolism to work. TIBTECH 11:360-367.

Bowman JP, Jimenez L, Rosario I, Hazen TC, Sayler GS (1993) Characterization of the methanotrophic bacterial community in a trichloroethylene-contaminated subsurface groundwater site. Appl Env Microbiol 59:2380-2387.

Caplan JA (1993) The worldwide bioremediation industry: prospects for profit. TIBTECH 11:320-323.

Ensley BD (1991) Biochemical diversity of trichloroethylene metabolism. Annu Rev Microbiol 45:283-299.

Feigel BJ, Knackmuss HJ (1988) Bacterial catabolism of sulfanilic acid via catechol-4-sulfonic acid. FEMS Microbiol Lett 55:113-118.

Feigel BJ, Knackmuss HJ (1991) Degradation of sulfanilic acid by a syntropic culture. Sulfanilic acid degradation to maleylacetic acid by *Pseudomonas paileronii* and *Agrobacterium radiobacter* mixed culture. Ind Waste Disposal M1:125.

Goszcynsky S, Paszcynski A, Pasti-Grigsby MB, Crawford RL, Crawford DL (1994) New pathway for degradation of sulfonated azo dyes by microbial peroxidases of *Phanerochaeta chrysosporium* and *Streptomyces chromofuscus*. J Bacteriol 176:1339-1347.

Hall JE, L'Hermite P, Newman PJ (1992) Treatment and use of sewage sludge and liquid agricultural wastes. ECSC-EEC-EAEC Brussels, ISBN 92-826-4142-2.

Hutchins SR, Biodegradation of monoaromatic hydrocarbons by aquifer microorganisms using oxygen, nitrate, or nitrous oxide as the terminal electron acceptor. Appl Env Microbiol 57:2403-2407.

Kovacs KL, Polyak B (1991) Hydrogenase reactions and utilization of hydrogen in biogas production and microbiological denitrification systems. Proc 4th IGT Symp, Colorado Springs, Chapter 5, pp 1-16.

Liu S, Suflita JM (1993) Ecology and evolution of microbial populations for bioremediation. TIBTECH 11:344-352.

Locher HH, Leirsinger T, Cook AM (1991) 4-toluene sulfonate methyl-monooxygenase from *Comamonas testosteroni* T-2: purification and some properties of the oxygenase component. J Bacteriol 173:3741-3748.

Lowe SE, Jain MK, Zeikus JG (1993) Biology, ecology, and biotechnological applications of anaerobic bacteria adapted to environmental stresses in temperature, pH, salinity, or substrates. Microbiol Rev 57:451-509.

Mohn WW, Tiedje JM (1992) Microbial reductive dehalogenation. Microbial Rev 56:482-507.

Murrell JC, Dalton H (1992) Methane and methanol utilizers. Plenum Press, New York, ISBN 0-306-43878-X.

Tsien HC, Brusseau GA, Brusseau RS, Hanson RS, Wackett L (1989) Biodegradation of trichloroethylene by *Methylosinus trichosporium* OB3b. Appl Env Microbiol 55:2960-2964.

Van Loosdrecht MCM, Heijnen SJ (1993) Biofilm bioreactors for waste-water treatment. TIBTECH 11:117-121.

BIOREMEDIATION OF FOUR FORMER SOVIET MILITARY BASES

IN HUNGARY: EXPERIENCE USEFUL FOR FUTURE DECISION-

MAKING

Tibor T. Sarlos and Károly Gondár
Comco Martech Hungary, a division of Comco Martech Europe AG
Patakhegyi út 83-85/C
1028 Budapest
Hungary

Abstract

Large quantities of soils were contaminated with various types of petroleum hydrocarbons at four former Soviet military bases in Transdanubian Hungary. The contamination resulted from approximately forty years of improper storage and handling of fuels and oils used in association with military operations at the four sites. These sites are a heliport at Székesfehérvár, and combat support bases in Veszprém, Tab, and Dombóvár. Much of the contamination, consisting of various components of kerosene, diesel oil, heating oil, gasoline, and lubricating oils and greases, was the result of uncontrolled releases from fuel and oil storage tanks, both underground and aboveground.

At the four sites scheduled for cleanup over a period of about three years and in the framework of several contracts with the Hungarian Ministry for Environmental Protection and Regional Policy (HMERP), approximately sixty fuel and oil storage tanks were removed, and 72,000 cubic meters of contaminated soils were excavated and biologically remediated.

In most cases, one growing season (April to September) was sufficient for the total petroleum hydrocarbon (TPH) concentration in the soil to reach the cleanup limit of 100 parts per million (ppm) which was the limit prescribed both contractually and by the local regulatory authorities in the two regions. In the case of some of the more contaminated, less air-porous soils at Veszprém and Tab, 1.5 to 2.0 entire growing seasons were required. Soils reached the cleanup limit and were certified as clean in several phases, spanning between October 1992 and December 1993; all of the soils were certified as clean in 1993. Subsequent to closure

NATO ASI Series, Partnership Sub-Series, 2. Environment – Vol. 1
Clean-up of Former Soviet Military Installations
Edited by R. C. Herndon et al.
© Springer-Verlag Berlin Heidelberg 1995

sampling and certification of the attainment of the cleanup level, soils were backfilled to their original locations, and the sites were turned back over to the Hungarian State Property Agency, or in the case of the Veszprem Kossuth Base, to its new owner, the Municipality of Veszprem.

This experience demonstrates that bioremediation using the landfarming technique is a cost-effective solution for soil remediation, and it works even with stringent cleanup limits. This technology is relatively inexpensive. Additional savings can be realized at vacated bases in Eastern Europe due to the lower cost of labor and the abundance of wide-open, paved surface areas which are ideal for constructing landfarming parcels at most military sites. Furthermore, the current work provides evidence that biostimulation can be effective under certain circumstances, resulting in further monetary savings, and should be seriously considered as a possible alternative to bioaugmentation.

Key Words: bioremediation, petroleum hydrocarbons, *ex situ*, bioaugmentation, biostimulation, cost-effective, soil remediation, Soviet military bases, Hungary, cleanup limit, landfarming

Background

The dramatic political changes that reshaped Eastern Europe in 1989 resulted in the evacuation of Soviet troops and hardware from hundreds of military installations throughout the region. The Soviet military had operated, with minimal regard for environmental preservation, as evidenced by large amounts of contaminated soils and groundwater discovered at the vacated sites. As contamination at an increasing number of former Soviet bases in Hungary was being discovered, Hungarian government officials acted immediately to mitigate potential health and environmental risks posed by the contamination. Due to the perceived urgency of the situation, as well as the limited financial resources available to the Hungarian Ministry for the Environmental and Regional Policy (HMERP), the decision was made to conduct preliminary assessments rapidly at a number of those former Soviet military sites suspected of being significantly polluted, and to begin remediation where necessary as quickly as possible.

Based on the preliminary assessments which were performed at vacated Soviet bases in Hungary, a variety of types of pollution were found, principally consisting of petroleum contaminated soils resulting from leaking underground storage tanks. Components of gasoline, diesel, and kerosene fuels were the principal contaminants, and total petroleum hydrocarbon (TPH) concentrations ranged up to 10,000 parts per million (ppm) were also found. As a result of the assessments, the HMERP initially ordered remediation of three military bases located in Veszprém, Székesfehérvár, and near Tab. Later, in a second round of base cleanup contracting, Comco Martech was awarded a fourth base, near Dombóvár, at which the objective was to localize the contamination in order to limit further environmental damage. At the four bases, a total of approximately sixty underground and aboveground fuel storage tanks were removed and disposed of and 72,000 m^3 of contaminated soils were excavated for biotreatment. In order to keep costs to a minimum, bioremediation by landfarming was selected as the method of treating the large quantity of contaminated soils. Hungarian labor, equipment and materials were used to the fullest extent possible. Regulatory complications associated with importing hydrocarbon degrading bacteria into Hungary and obtaining approval to use them in the field were encountered; hence, bacterial cultures already approved for use within Hungary for treatment of hazardous wastes were identified for use.

Bioremediation of the soils excavated was completed in accordance with the requirements prescribed by relevant local regulatory agencies, and the sites have been restored and turned over to their previous or new owners. The cleanup limit of 100 ppm TPH was attained at different times for different lifts of soil at the four bases; cleanup times ranged from less than one growing season (April to September) to two growing seasons.

Site Locations/Descriptions

All four sites are located in Transdanubian Hungary, i.e., generally west of the Danube River. Veszprém, Székesfehérvár and Tab are about 20, 30, and 50 kilometers north, northeast, and south of Lake Balaton, respectively. Dombóvár is approximately 80 kilometers south of Lake Balaton, in the direction of Pécs.

The former "Kossuth Lajos" Soviet military base in Veszprem is situated in an area of karst terrain, containing primarily clay-type soils mixed with some gravel at shallower depths. Sandy soils are predominant at the former Soviet heliport in Székesfehérvár, while soils at the "Hunyadi Janos" military base near Tab in Somogy County can be characterized as silty clays and clayey silts. Sand and silt deposited by the Kapos River make up the soils at the former "Tolbuchin" base between Dombóvár and Kaposszekcsô.

The sites at Tab, Veszprém, and Dombóvár were established as Hungarian military bases before the 1950s, but were taken over by Soviet troops as they strengthened forces here after stamping out the Hungarian revolt against Soviet domination in 1956. The Taci ut heliport at Székesfehérvár was established for the Soviet military around 1960. All four bases were vacated by the Soviet Army in mid-1990.

Type and Extent of Contamination

The soils treated were contaminated with diesel fuel, gasoline, lubricating oil and grease and heating oil at Veszprém and Tab, kerosene at Székesfehérvár, and diesel fuel and lubricating oils at Dombóvár. Some of the contamination was recent, but most of it had been subject to atmospheric weathering for up to 40 years. Table 1 illustrates the basic characteristics of contamination in soils at each of the four bases.

Table 1

Site	Concentration Range TPH (ppm)	Concentration Average TPH (ppm)	Type
Veszprém	0 - 7,000	2,000	diesel fuel, heating oil, waste oil
Tab	0 - 2,500	1,500	diesel fuel, gasoline, waste oil
Székesfehérvár	0 - 10,000	500	kerosene
Dombóvár	0 - 3,000	650	diesel fuel, oil

Target Cleanup Levels

The target cleanup levels, contractually stipulated in 1991 and strictly adhered to by the regulatory agencies for the duration of the remediation program, were:

- Total Petroleum Hydrocarbons (TPH): 100 parts per million; and
- Total Polyaromatic Hydrocarbons (PAH): 1 part per million.

These limit values are based on the Hungarian technical guideline MI-08-1735-1990 (see Table 17 in MI-08-1735-1990) which prescribes the maximum contaminant concentrations allowable in agricultural soils following sewage sludge applications. These limits were not, therefore, entirely appropriate for the purpose of this remediation project; however, for lack of a more applicable standard, they were stipulated as permit conditions by the relevant local regulatory agencies. Because the guidance mentioned is designed to regulate contaminant concentrations in arable land, the cleanup limits are "health conservative," or conservative from the point of view of environmental protection and public health. However, such stringent cleanup limit values may have been overly rigorous for some of the bases if, for example, industrial activities will take place at these sites in the future. Financial resources might have been saved by taking site-specific factors, such as future land use and expected exposure levels for potential receptors, into consideration in setting cleanup limits.

Methodology

Bioremediation using landfarming methods was selected for the remediation of contaminated soils at the four military bases. In general, bioremediation consists of using bacteria or other biota to metabolize and remove contaminants from soil or water. Bioremediation has been applied successfully to remove petroleum hydrocarbons from contaminated soils (Loehr, 1984; Raymond et al., 1976; Huddleston et al., 1984). As opposed to other methods of soil treatment which typically transfer the contaminants from one part of the environment to another, bioremediation utilizes natural processes which reduce toxic contaminants to innocuous components. In the case of oil or fuel-contaminated soil, under ideal conditions the hydrocarbons are converted by soil microorganisms to carbon dioxide, water and biomass (bacterial cells). Such processes are carried out spontaneously by bacteria indigenous to the impacted soils under many conditions; however, the natural degradation process can be significantly accelerated by stimulating the growth of these organisms (biostimulation) or employing additional populations of petroleum degrading microorganisms (bioaugmentation).

Both biostimulation and bioaugmentation were employed at these former Soviet military bases. The essential component of the landfarming methodology is stimulating the growth and productivity of the microorganisms by carefully controlling the environmental conditions in the biopad. Numerous factors are known to affect both the kinetics and the extent of hydrocarbon removal from contaminated soil. These include factors such as pH, temperature, moisture, aeration, and nutrient levels (e.g., nitrogen and phosphorus), contaminant characteristics such as molecular structure and toxicity to microbes, and the ecology of the microbial populations present in the soil (Huddleston et al., 1984; Atlas, 1991; Bartha, 1986).

The landfarming approach usually begins by the construction of bioremediation 'lifts' or 'pads' which are used to contain the soils requiring treatment. The pads consist of a flat area lined with either a relatively impermeable liner, or with extensive concrete covering to prevent the migration of pollutants from the contaminated to the underlying clean soils. The treatment area is then bermed to control runoff in the event of heavy precipitation. The former military bases were ideal for landfarming since numerous large, open areas paved with asphalt or concrete surfaces exist on site.

The contaminated soils were excavated and spread out on concrete or asphalt areas at each of the four sites, and amended with fertilizers, water, and soil bulking agents to enhance the natural biodegradation process. At three of the four bases, the indigenous soil microbial population in contaminated soils was bioaugmented with petroleum-degrading bacteria, while at Székesfehérvár, biostimulation was sufficient. By applying this latter method, some of the money saved by not buying commercial microorganisms is usually spent on treatability studies in the laboratory to identify the site-specific conditions optimal to microbial growth and degradation of hydrocarbons. Biostimulation is becoming more common in the United States, and is currently viewed by many environmental professionals as "state-of-the-art".

Analytical Methods

From the outset of this program, gas chromatography (GC) was utilized to analyze soil samples for TPH. At various stages, the local regulatory agencies performed soil analyses by either gravimetric methods or ultraviolet infrared spectroscopy in order to monitor the progress of the bioremediation. At the time, these were the only analytical methods approved for use in Hungary for the determination of TPH concentrations in soil. However, for internal progress tracking purposes, samples were analyzed by GC throughout the project.

Toward the end of the project in 1993, the Hungarian regulatory agencies in charge of supervision of the remediation were able to accept results obtained by GC for official determination of TPH concentrations in soil, and the final results were analyzed by GC. Apparently, numerous questions and concerns regarding the accuracy of the methods other than GC had arisen to prompt decision-makers to compare the accuracy of various methods and to rethink their reasoning about which methods could be officially accepted. This is one example of a positive development in the Hungarian environmental remediation industry which resulted partly from experiences associated with the remediation of the former Soviet military bases.

Results and Discussion

The amount of time it took to degrade the TPHs to the cleanup limit concentration of 100 ppm varied between four months in the case of one lightly contaminated biopad at the Tab Base to two growing seasons (April through September), which was the time necessary to treat about 50% of the total volume of soils at the Tab Base. Most bacteria that carry out bioremediation processes are mesophiles ("middle lovers") and are most active in the temperature range of 18 to 30 degrees centigrade. Significantly higher or lower temperatures will limit their activity. Therefore, it is expected that significant biodegradation only occurred at the bases between the months of April and September; hence times to reach the cleanup limit are counted in number of growing seasons. Table 2 illustrates the amount of time necessary to reduce TPH concentrations to the cleanup level for the various biopads at the four bases.

Table 2

Site	Biopad	Initial Average Conc. (ppm)	Start	Finish	No. of Growing Seasons
Székesfehérvár		500	April, 1992	July, 1992	1.0
Veszprém	1	1,460	April, 1992	June, 1993	1.5
Phase 1	2	1,570	April 1992	June, 1993	1.5
	3	1,200	April, 1992	October, 1992	1.0
	4	2,510	April, 1992	June, 1993	1.5
	5	1,990	April, 1992	June, 1993	1.5
	6	960	April, 1992	June, 1993	1.5
	7	540	April, 1992	October, 1992	1.0
	8	1,150	April, 1992	October, 1992	1.0
	9	790	April, 1992	June, 1993	1.5
	10	1,520	April, 1992	July, 1993	1.5
	11	860	April, 1992	July, 1993	1.5
Phase 2	VII	1,530	April, 1993	October, 1993	1.5
	VIII	970	April, 1993	October, 1993	1.0
	XII	3,580	April, 1993	November, 1993	1.0
Tab	1	530	April, 1992	November, 1992	1.0
Phase 1	2	250	April, 1992	November, 1992	1.0
	3	760	April, 1992	November, 1992	1.0
	4	720	April, 1992	October, 1993	2.0
	5	1,270	April, 1992	October, 1993	2.0
	6	890	April, 1992	October, 1993	2.0
	7A	510	April, 1992	December, 1993	2.0
	7B	510	April, 1992	November, 1992	1.0
	8	2,000	April, 1992	October, 1993	2.0
Phase 2	II	650	April, 1993	October, 1993	1.0
	IX	240	April, 1993	July, 1993	1.0
Dombóvár	1	580	October, 1992	October, 1993	1.0
	2	1,030	October, 1992	December, 1993	1.0

Székesfehérvár

At the Székesfehérvár heliport, one growing season was sufficient to reduce TPH levels to the cleanup limit, utilizing only the indigenous bacterial population. This can be explained by the soil type (primarily sand) and by the fact that the contamination was comprised of light-fraction hydrocarbons (mostly kerosene). This is a prime example of a good site for biostimulation, because soil and contamination characteristics are ideal for bioremediation and the contamination had been there long enough to have propagated a healthy indigenous population of hydrocarbon-degrading soil microbes.

Veszprém

The soil remediation at the Veszprém site was contracted in two phases, and because of the clay-type soils and the heavier nature of the contamination, a bacterial inoculum was applied in both phases. With the clays at Veszprém, it was relatively more difficult to maintain the air-filled porosity needed to supply the bacteria with the required oxygen than at the other bases. Bulking agents were applied to improve the potential of oxygen to reach the soil microorganisms. Despite this difficulty, in the first phase of the work, approximately 80% of the soils were clean after one growing season of treatment, while the other 20% needed one-and-a-half growing seasons. In the second phase of treatment, all the soils were clean after one growing season, presumably resulting from intensified soil condition adjustment activities as well as to the application of a bacterial starter culture which was more efficient at degrading the compounds present at the site.

Tab

During the first phase of remediation activities at the Tab Base, approximately 35% of the soils were clean after one growing season, while the remaining 65% needed two growing seasons. This can be largely attributed to the fact that, although the soils at the Tab Base were more easily managed with respect to aeration and moisture control. While the initial TPH concentrations were lower than those at Veszprém, the contamination largely consisted of heavier-fraction hydrocarbons such as lubricating oils and greases. The second phase of soil remediation at this site was completed in one growing season or less, partly because of lighter initial contamination levels, and partly because more care was taken to apply a bacterial starter culture specifically suitable for degradation of the compounds present at the site.

Dombóvár

The soils at Dombóvár were brought to the cleanup limit concentration in one growing season. This is likely due to the relatively low level of initial contamination, the nature of the pollutants (primarily diesel fuel) as well as the structure of the soils, which were comprised of sands and silts. This coarse-grained type of soil is much more conducive to effective aeration than the soils found at the Tab or Veszprém sites, which contain partly or mostly fine clays.

Based on the experience gained at the four bases, it can be concluded that the time necessary for bioremediation was influenced more by the type of the contamination than the actual magnitude of the TPH concentration in the soils. The speed of the process also clearly depends on soil type, though to a lesser extent, as this can be modified by the addition of bulking or other amendments which improve the aeration and moisture maintenance characteristics of the soil. Since the biopads were outdoors and exposed to the elements, climatic conditions at each site also had an influence on the duration of bioremediation. For example, at all four bases it was very difficult to maintain the optimal moisture content for biodegradation during certain rainy or windy periods. This was especially true at Veszprém as the clay particles stick together tightly when the soils are too moist; on the other hand, when exposed to sun and wind, the soil dries into big, hard chunks.

Costs

The *ex situ* bioremediation was completed at a cost of approximately 1,600 HUF/m^3 of soil (HUF = Hungarian Forints). This is approximately 35-40% less than the cost of a similar project in the United States. Most of these savings can be attributed to the relatively low cost of labor in Hungary. Laboratory analyses, a significant portion of environmental project costs (20-50%), are also cheaper in Hungary than in the United States, but again, this can mostly be attributed to the current cheaper cost of labor in Eastern Europe.

Although site-specific costs varied greatly between different bases, the 1,600 HUF/m^3 figure is a good indication of the current average cost of *ex situ* bioremediation for the type of contamination and soils encountered at the four military bases. Addressing the sites separately, remediation costs varied anywhere from 800 HUF/m^3 for lightly contaminated, more easily bioremediated soils, to 3,000 HUF/ m^3 for some of the more heavily contaminated soils which are less conducive to bioremediation.

Factors Influencing Cost

Biodegradation of petroleum hydrocarbons is a process which will eventually take place naturally in the environment if its main requisites are satisfied (e.g., oxygen, food source, nutrients, moisture). The purpose of the applied bioremediation technology is simply to expedite this natural process by creating optimal conditions for the phenomenon to occur. Costs, therefore, are greatly dependent on the extent to which an instance of contamination (soil and pollutants in the soil) lends itself to manipulation with respect to the conditions required for biodegradation. That is, costs depend not only on the level of contamination and how biodegradable the compounds present are, but how amenable the soil is to aeration (the addition of oxygen), moisture control, nutrient level maintenance, etc.

As mentioned previously, many factors affect the time necessary for the microorganisms to degrade the substrate into its components. The more resistant the characteristics of a particular instance of contamination are to bioremediation (e.g., the heavier the TPHs, the smaller the soil particle size, the more severe the weather conditions) the longer the process. Since soil conditions must be monitored and adjusted continuously as long as the bioremediation is in progress, the more time it takes, the greater operational costs are incurred, and in turn, the higher the total project cost will be. Clearly, one conclusion is that the total project cost (and cost per m^3) are dependent on the temporal duration of biodegradation.

In summary, the main factors affecting the level of difficulty, and therefore the cost of TPH remediation by *ex-situ* landfarming observed in this experience were: initial magnitude of TPH-contamination, type of petroleum product present, soil type, and site-specific weather conditions.

If time is an important factor, as it is with many environmental restoration projects; remediation deadlines are often stipulated by the client or the regulatory agency. If only a relatively short time is available to complete bioremediation, project costs will become significantly higher when additional bacteria and adjustments to soil characteristics are necessary to expedite the process. Time requirements, therefore, may cause significant incremental increases in total project cost. Conversely, significant cost savings may be achieved with a more flexible cleanup deadline.

If soils are heavy and expensive to transport, the distance between the contaminated areas and suitable open surfaces for landfarming can also significantly affect the cost of landfarming. Hence, significant cost savings are gained because of the ample availability of paved surface for constructing biopads at most former military sites.

Conclusions Useful for Future Decision-Making

The following conclusions, gleaned from the experience remediating 72,000 m^3 of contaminated soils at former Soviet military bases in Hungary, may be useful for future decision-making regarding the selection of innovative, cost-effective technologies for the remediation of these sites or other sites similarly polluted throughout the region.

1. Compared with other technologies, bioremediation is an inexpensive method for the remediation of petroleum product-contaminated soils. It is a natural process which eliminates the contamination rather than transferring it from one part of the environment to another.

2. Both *ex-situ* and *in-situ* bioremediation may be feasible, depending on the soil type and other site-specific circumstances. However, military bases are generally suitable, if not ideal, sites at which to apply *ex-situ* bioremediation because of the availability of open asphalt or concrete surfaces on which contaminated soils may be spread out for landfarming. The added costs of excavation and backfilling are expected to be outweighed by economic and time savings resulting from the increased ease of monitoring and controlling conditions conducive to efficient biodegradation of the target compounds.

3. For soil bioremediation projects in which financial constraints are more important than time constraints, a biostimulation approach should be considered before bioaugmentation. Substantial savings can be realized by stimulating the indigenous microbial population rather than by augmenting the soil with a selected culture of microorganisms. Some of the money saved by not purchasing expensive commercial microorganisms can be spent on increased monitoring and laboratory studies regarding site-specific biotreatability, as well as determining and achieving optimal values for field parameters, which are critical for successful, rapid biodegradation.

4. Careful consideration should be given to prescribing cleanup limits for remediation of soils and groundwater at contaminated sites, both military and other based on a risk assessment for the site. Whenever possible, cleanup limits should be set on an individual site basis, and should take into account factors which influence the realistic risk the site may pose to public health or the environment. Factors to be considered when developing cleanup levels include: land use after remediation (which partly defines the potentially exposed population), proximity to sensitive receptors, potential human exposure pathways, and the toxicity of the chemicals. Such an approach will minimize the likelihood that valuable resources are used to clean up sites to an unnecessary level of cleanliness, while other sites at which contamination may pose a risk to public health, for example, go untouched due to lack of resources.

5. Moreover, in order to properly establish environmental cleanup priorities, sites should be ranked by their potential to pose health or environmental risks at the outset of government programs involving remediation of contaminated sites. Site-specific factors such as those listed in conclusion #4 should be considered. After comprehensive environmental assessments have been conducted for each site, a ranking can be done with methods commonly used in the field of health risk assessment in the United States. There are risk assessment methods specifically developed for ranking or screening (e.g., separation of high risk versus low risk

sites) contaminated sites. Such screening methods include the employment of exposure assumptions and equations, and risk factors for carcinogenic and non-carcinogenic substances based on chemical toxicity data, exposure levels and other critical assumptions regarding the site.

In summary, Comco Martech's experience with the cleaning of petroleum contaminated soils at former Soviet military bases in Hungary demonstrates that bioremediation using the landfarming technique is a cost-effective solution for soil remediation, and is successful, even with stringent cleanup limits. This technology is comparatively inexpensive and additional savings can be realized specifically at former military bases in Eastern Europe due to the current lower cost of labor and the abundance of wide-open paved surfaces which are ideal for constructing landfarming parcels. Furthermore, this work provides evidence that biostimulation can be effective under certain circumstances, results in further monetary savings, and should be seriously considered as a possible alternative to bioaugmentation. Finally, the experience of having worked with rigorous cleanup levels leads to the recommendation that site-specific factors such as future land use, potential exposure, toxicity and attendant health risks be considered in setting site-specific cleanup levels.

References

Atlas, R.M. 1991. Bioremediation of Fossil Fuel Contaminated Soils. In: *In Situ Bioreclamation -- Applications and Investigations for Hydrocarbon and Contaminated Soil Remediation*, pp. 14-32. (Hinchee, R.E. and Olfenbuttel, R.F., Eds.), Boston, Butterworth--Heinemann.

Bartha, R. 1986. Biotechnology of petroleum pollutant biodegradation. Microbial Ecol. 12, 155-172.

Loehr, R.C. 1984. Land treatment as a waste management technology. In: Land Treatment -- A Hazardous Waste Treatment Alternative, pp. 7-17, (Loehr, R.C., and Malina, J.F., Eds.) Water Resources Symposium Number 13, University of Texas at Austin.

Huddleston, R.L., Bleckmann, C.A., and Wolfe, J.R. 1984. Land treatment biological degradation processes. In: Land Treatment -- A Hazardous Waste Treatment Alternative, pp. 41-61, (Loehr, R.C., and Malina, J.F., Eds.),Water Resources Symposium Number 13, University of Texas at Austin.

Raymond, R.L., Hudson, J.O., and Jamison, V.W. 1976. Oil degradation in soil. Appl. Environ. Microbiol. 31(4), pp. 522-535.

OVERVIEW OF BIOVENTING TECHNOLOGY FOR THE REMEDIATION OF PETROLEUM HYDROCARBON CONTAMINATION

Catherine M. Vogel
AL/EQW-OL
139 Barnes Drive, Suite 2
Tyndall AFB, FL 32403-5323
USA

Introduction

Bioventing is an innovative technology for the treatment of vadose zone soils contaminated by a wide variety of petroleum distillate fuels. Bioventing stimulates the *in situ* biodegradation of the hydrocarbon contaminants by providing oxygen to the native soil microorganisms. This technology is a variant of soil venting or soil vapor extraction (SVE) technology. The bioventing process utilizes a much lower air flow rate than typically is used in SVE. These lower air flow rates are designed to provide adequate oxygen to sustain microbial degradation while minimizing hydrocarbon loss through volatilization.

The least expensive and simplest bioventing design is *controlled air injection* into the contaminated zone. However, air extraction may be necessary due to site specific conditions, such as proximity to subsurface structures or excessive surface emissions. At sites with these conditions, alternatives to direct air injection can be considered. These include isolation of subsurface structures to prevent migration of the contaminated vapors, or air extraction combined with reinjection of the contaminated off-gas.

Field Testing

A controlled pilot-scale study of bioventing was conducted at Tyndall Air Force Base (AFB), Florida. The objectives of this study were to determine:

1. How air flow rate affects the amount of hydrocarbon removal by biodegradation and volatilization;

NATO ASI Series, Partnership Sub-Series, 2. Environment – Vol. 1
Clean-up of Former Soviet Military Installations
Edited by R. C. Herndon et al.
© Springer-Verlag Berlin Heidelberg 1995

2.   If nutrient and/or moisture additional, coupled with air flow, would enhance biodegradation rates; and

3.   If contaminated off-gas could be effectively treated by injecting the air stream into clean soil.

The test area was located at a site previously used for the storage of JP-4 jet fuel, and where fuel storage and transfer activities are suspected to have resulted in soil and groundwater contamination. The soils consisted of fine-to-medium grained quartz sand. The depth to groundwater was approximately 4 feet.

In order to conduct the bioventing experiments, enclosed plots of contaminated soil were constructed. This design allowed for a rigorous mass balance to be achieved on total petroleum hydrocarbons and oxygen. The site was dewatered to expose a larger volume of the contaminated smear zone to the bioventing process. Through dewatering, the groundwater table was maintained at approximately 6-7 feet.

Results from the Tyndall AFB study are presented in Miller et al. 1991. Nutrient and moisture addition had no statistically significant effect on the hydrocarbon biodegradation rates. However, *in situ* biodegradation rates did correlate well with *in situ* soil temperatures as predicted by the van't Hoff-Arrhenius equation. The amount of hydrocarbon removal attributed to *in situ* biodegradation was increased to 85 percent due to proper management of the venting air flow rates. Hydrocarbon contaminated air streams could be treated if injected into clean soil and if adequate retention time was provided.

In order to transfer effectively the knowledge gained to-date regarding bioventing technology, the U.S. Air Force Center for Environmental Excellence (AFCEE) began the "Bioventing Initiative" in 1992. This initiative consists of testing the feasibility of bioventing at over 135 Air Force fuel contamination sites. The objectives of the initiative are:

- to document the feasibility of bioventing for petroleum hydrocarbon contamination under a variety of climatic, soil, and contaminant conditions;

- to promote regulatory and public acceptance of this technology;

- to generate a Bioventing Principles of Practice Manual from the large data set collected; and

- to begin remediating the 135 contaminated Air Force sites using this simple and inexpensive technology.

The contaminated sites used in the initiative were chosen based on the level of hydrocarbon contamination, absence of any floating product, and the absence of any chlorinated solvent co-contamination. Sites span a wide range of climatic and hydrogeologic conditions.

A bioventing protocol was developed as the first step of the initiative (AFCEE, 1992). All testing at the 135 contaminated sites followed the guidance in the protocol. Testing at each site includes a soil permeability test and an *in situ* respiration test. In order for bioventing to be successful, the soil must be permeable enough to allow air to flow through it and microbial activity must be measurable under aerobic conditions. If positive results are obtained from these two tests, a pilot-scale bioventing system is installed and allowed to operate for one year. Results from more than 60 of the 135 test sites have been compiled and presented in Miller et al. 1993.

Enhancements to the Bioventing Process

Additional field studies are being conducted by the U.S. Air Force, U.S. Environmental Protection Agency (EPA) and others to expand the range of sites where bioventing could be applied. A three-year, pilot-scale field study is ongoing at Eielson AFB, Alaska, to determine the effectiveness of bioventing used in conjunction with soil-warming techniques. This study is sponsored by the U.S. Air Force and the U.S. EPA. The objective is to determine if the combination of soil aeration plus soil warming can stimulate the *in situ* biodegradation of petroleum hydrocarbon contaminants in a sub-arctic environment. Three methods of soil warming are being tested in conjunction with aeration through air injection: passive soil warming (surface insulation); active warming - infiltration of heated water through the vadose zone; and surface warming via buried heat tape.

Preliminary results are very encouraging. Bioventing is stimulating *in situ* biodegradation and the soil warming methods are increasing *in situ* soil temperatures, thereby increasing *in situ* biodegradation rates. Details of the field experiments and interim results are presented in Leeson et al. 1993. An economic analysis of the field data will be completed at the end of the study (in December of, 1994) to determine the cost-effectiveness of applying bioventing with soil warming for remediation of fuel contamination in a sub-arctic environment.

A large-scale bioventing study is also being conducted at F.E. Warren AFB, Wyoming to test enhancements to the bioventing process. Field experiments are being conducted to determine the effect of pulsed air injection and pure oxygen pulsing on hydrocarbon biodegradation rates. This study is being conducted at an abandoned fire fighting training facility primarily contaminated with petroleum hydrocarbons.

Bioventing combined with a groundwater treatment technology is being evaluated at a fuel spill contamination site at Tyndall AFB, Florida. The bioventing concept was modified to incorporate an in-well sparging groundwater recirculation system. Results are not yet available from the studies at F.E. Warren AFB, Wyoming or Tyndall AFB, Florida.

Most of the bioventing work done to-date has been conducted at sites contaminated with gasoline, diesel fuels, and aviation fuels (JP-4 and JP-5). To expand the range of contaminants that bioventing can effectively remediate, the U.S. EPA is conducting a three-year study to evaluate the potential of bioventing to remediate soils contaminated with polycyclic aromatic hydrocarbons (PAHs). Field system design and preliminary results are described in McCauley et al. 1994.

Benefits of Bioventing

Bioventing is an extremely cost-effective technology to remediate unsaturated zone soils contaminated with petroleum hydrocarbons. Costs in the range of $10-60 (USD) per cubic yard of contaminated soil have been documented. At large contamination sites, with over 10,000 cubic yards of contaminated soil, costs of less than $10 (USD) per cubic yard have been achieved. These costs assume that no off-gas treatment will be required. This is easily achievable with proper system design and operation. Research and development currently being conducted will offer further optimizations to the bioventing process, as well as an expanded range of contaminants for which the technology can effectively remediate.

References

Air Force Center for Environmental Excellence, "Test Plan and Technical Protocol for a Field Treatability Test for Bioventing", National Technical Information System Number PB 93-209146, May, 92.

Leeson, A., R.E. Hinchee, J. Kittel, G. Sayles, C.M. Vogel, and R.N. Miller. "Optimizing Bioventing in Shallow Vadose Zones and Cold Climates", Hydrological Sciences, 38:283-295, 1993.

McCauley, P.T., R.C. Brenner, and B.C. Alleman. "Bioventing Soils Contaminated with Wood Preservatives", Proceedings, US EPA Symposium on Bioremediation of Hazardous Wastes: Research, Development, and Field Applications, San Francisco, CA, June, 1994.

Miller, R.N., C.M. Vogel, and R.E. Hinchee. "A Field-Scale Investigation of Petroleum Hydrocarbon Degradation in the Vadose Zone Enhanced by Soil Venting at Tyndall AFB, FL", In: (R.E. Hinchee and R.F. Olfenbuttel, eds.) In Situ Bioreclamation, pp 283-302, 1991.

Miller, R.N., D.C. Downey, V.A. Carmen, R.E. Hinchee, and A. Leeson, "A Summary of Bioventing Performance at Multiple Air Force Sites", Proceedings, Petroleum Hydrocarbons and Organic Chemicals in Groundwater: Prevention, Detection, and Restoration, Houston, TX, pp 397-411, 1993.

RESULTS OF REMEDIAL TECHNOLOGIES APPLIED AT VÁC-MÁRIAUDVAR, A FORMER SOVIET MILITARY INSTALLATION

Judit Tóth
Dekosta-Biokör
Építész u. 40-44.
H-1116 Budapest
Hungary

Introduction

The Soviet Army (SA) caused considerable damages to a large number of military installations formerly used as airports, barracks, military ammunition plants and fuel storage bases in Hungary. Preliminary assessments have revealed massive hydrocarbon contamination of soil and groundwater at most of these locations.

The Ministry for Environment and Regional Policy issued tenders for the remediation of thirteen of these abandoned military sites. One of these sites was the Vác-Máriaudvar military base, for which the remedial project was offered to and accepted by Dekosta-Biokör. The project started in August of 1992.

Site Description

The site is situated 30 km north of Budapest. Its covers an area of 16 hectares, and from 1938 to 1991 it was used as a fuel storage base. The installation was built in 1938 by the Germans to service a Wermacht airbase, which was subsequently not built. The center of the installation consists of three pumping houses each of which is surrounded by 20 underground tanks of 50 m^3 volume. The enameled tanks and the attached pipes are in good condition even now. The Germans used the piping network to transport gasoline into the tank systems.

After 1945, when the property came under the control of the SA, the piping network was no longer used. The Soviets used tank trucks to transfer the fuel from railcars to fuel storage tanks throughout the area. Spillage occurred as a result of these activities. Later the Soviets placed about 200 above-ground storage tanks along the northeast border of the site. These

NATO ASI Series, Partnership Sub-Series, 2. Environment – Vol. 1
Clean-up of Former Soviet Military Installations
Edited by R. C. Herndon et al.
© Springer-Verlag Berlin Heidelberg 1995

tanks of various size contained 10-50 m^3 of diesel fuel or lubricating oils. The location had been used as a reserve fuel supply without significant traffic, yet at one time as much as 10,000 m^3 of oil were stored here. When the Soviets departed the site, most of the smaller tanks (between 10-25 m^3 volume) were removed and sold. The majority of the present contamination occurred at this time. More than a hundred tanks remain in place. While the fuel in those tanks was removed, the tanks were not cleaned or otherwise decommissioned.

Local Conditions and Remedial Status

The property is situated in the old river bed of the Danube. The soil consists of middle to coarse sand from the surface to a depth of 3 m. The zone from 3-5 m below ground surface, is a sandy silt layer interbedded with clay. Below 5 m is a coarse sand and gravel deposit which contains groundwater. The groundwater level is at 7-8 m below ground surface. A confining layer at 12-13 m is a clay zone of the Oligocene Epoch.

During the course of the investigation, borings were made at 32 locations throughout the entire area to the depth of the groundwater. Analysis of soil and water samples revealed three different types of contamination:

Surface contamination of soil

Surface soil contamination typically was detected at 0-0.3 m, but sometimes at a maximum of 1 m. The characteristic pollutant is lubricating oil in the vicinity of the railway transfer points, the lubrication oil tanks, and proximal to the motor vehicle service area.

The concentration of contamination is highly variable, generally about 0.2-5 g kg^{-1}, but around the oil tanks it was 5-30 g kg^{-1}, and at the motor vehicle service area it was 30-100 g kg^{-1}.

Underground diesel fuel contamination

Diesel fuel contamination could have originated in part from the leakage of tanks; however, it is believed that the bulk of the contamination occurred when the diesel tanks were removed during the departure of the SA, when the remaining fuel was discharged into the holes. In the sandy soil the vertical spread of pollution at some places reached 5 m; having a concentration in the soil of 1-10 g kg^{-1}.

Underground gasoline contamination

While the presence of diesel and lubricating oil contamination was readily detectable on the surface, the area surrounding the gasoline tanks appeared to be without noticeable contamination.

Initially, the soil pollution was indicated only by the dissolved gasoline in the groundwater. Based on monitoring well data, the contamination was concentrated in the vicinity of two pumping houses. To determine if the source of contamination still existed, a soil gas survey was performed for the area around the tanks belonging to these two pumping houses. Gas chromatographic analysis of the soil gas survey revealed concentrated levels of gasoline $(5,000-25,000$ mg m$^{-3})$ in the vicinity of the eight tanks. It is suspected that the high contamination in this area originated not from leakage, but rather from careless management practices. Upon inspection, all tanks and pipelines proved to be pressure-tight and intact.

Analysis of soil samples verified that the gasoline contamination occurred below 4 m and was concentrated above the saturated zone. Gasoline concentration was found in the range of 5-7 g kg^{-1} and, therefore, it was concluded that soil contamination represented a significant source of groundwater contamination. Free phase hydrocarbon on the water table was not found, but the concentration of dissolved components was 1-5 ppm in a significant radius. In the center, the contamination reached values exceeding 50 ppm (Figure 1).

Selection of Remedial Actions

In accordance with the contractual agreement, the following two goals were to be achieved in the first phase of remedial action:

- to stop the off-site migration of groundwater pollution;
- to reduce the amount of residual soil contamination (the source of groundwater contamination).

To achieve these two goals, a pumping and treatment system was selected to remove the strippable components from the groundwater, and a soil venting technology was selected for the removal of the contaminant source in the soil. In the gasoline contaminated area, the remedial system was composed of the following parts:

- 3 extraction pumping wells were placed at the centers of contamination and the pumped groundwater, which was mainly contaminated with volatile components, was cleaned in a stripping tower;
- as part of the soil venting system, five air extraction wells were installed and connected to four blowers; and

Fig. 1. Distribution of ground-water contamination

REMEDIAL ACTION AT VÁC-MÁRIAUDVAR, A FORMER SOVIET MILITARY INSTALLATION
August, 1992

Distribution of ground-water contamination

dissolved gasoline
(mg/dm³)

0-1
1-2
2-4
4-6
6-8
8-10
10-15
15-20
20-30
30-50
50-

Made by the GEOCOMP Ltd.
based on the DEKOSTA-BIOKÖR Ltd. survey.

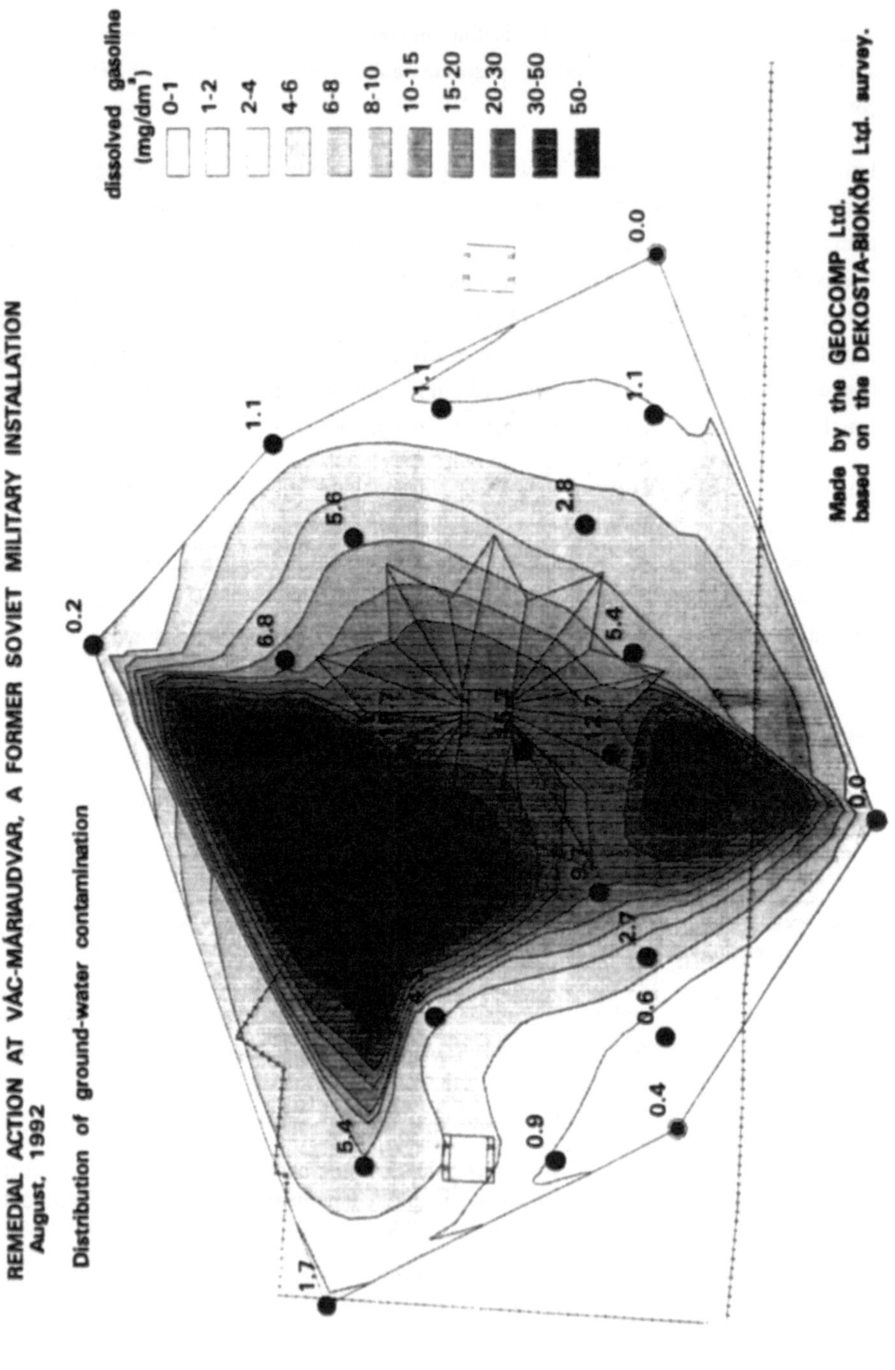

- an activated carbon filter was used to remove the volatile components from air extracted from the soil, and to clean the air from the stripping tower.

Soils were excavated in the area contaminated mainly with non-volatile organic components. Excavation was performed where surface contamination was more than 5 g kg^{-1}, as well as when diesel oil contamination attained or exceeded certain depths. The excavated soil was treated on-site, while the surface contamination was treated using a land farming process.

Treatment Techniques

On-site biological soil treatment

Three thousand cubic meters of excavated soil were treated on-site. A quarter of that volume contained lubricating oils, while the main pollutant was diesel oil.

The first goal in treating the soil was to mix and to loosen the heavily contaminated soil with the mildly contaminated soil. An equalized nutrient supply was provided by adding organic manure and fertilizer. The soil was mixed with nutrients and inoculated with oil-degrading microorganisms, cultivated on-site. The soil was piled up to a height of 1.5 m on a plastic sheet with drains. Aeration was accomplished by pulling air through built-in ventilation pipes.

The total volume of contaminated soil was treated in 5 piles. The volume of each pile was approximately 600 m^3. Initial mean contamination values of the piles varied. During the nine months of operation time, continuing through the winter period, the hydrocarbon contamination decreased, eventually to an acceptable limit. In June of 1993, the piles were backfilled into the excavated site. Based on sampling and control measurements, the treated soil quality satisfied the remediation standards. The efficiency of degradation was 90% (Figure 2). The total hydrocarbon residual was 570 ppm. While the diesel oil components were degraded, the residual contamination was from the lubrication fraction. Considering the future use of the property, the results were better than expected.

Land farming process

Where the surface contamination was not significantly greater than 5 g kg^{-1} or deeper than 0.5 m, the soil was not removed. The treatment was performed at four different sections of the property, including the areas surrounding the oil tanks and the railway tracks.

At these sections, a turning plow and tiller was used to mix the soil and increase the oxygen level. Nitrogen and phosphorus fertilizers were added, and an oil-degrading inoculum spray was applied. Since it was autumn, rye grass was planted to loosen the soil texture and promote water balance, instead of using *papilionaceae* (which has proven to be very useful in enhancing the speed of oil degradation). In the spring after the harvesting of the rye grass, the whole operation was repeated and finally alfalfa was sown. By the end of August of 1993, the target concentration limit of 300 ppm was achieved (Figure 3).

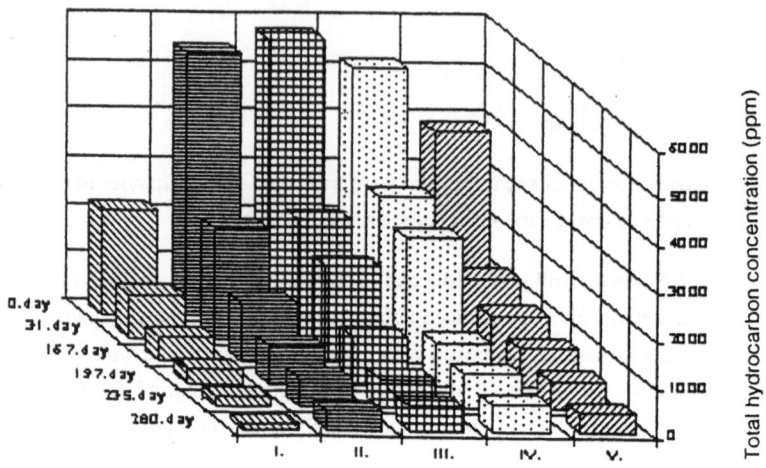

Figure 2. Hydrocarbon degradation in the on-site piles.

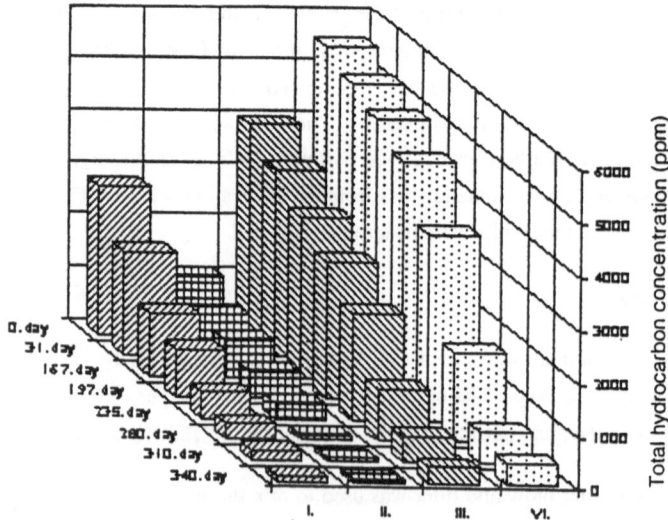

Figure 3. Hydrocarbon degradation in the land farming process.

Groundwater pumping and stripping

The regional groundwater flow is from east to west, but the local flow direction is influenced by the old river bed of the Danube. Based on this influence, it was necessary to increase the number of monitoring wells. The extraction well locations were determined on the basis of water quality and flow analysis.

At a pumping rate of 1.2 liters sec^{-1}, each well was shown to produce a radius of influence sufficiently large to contain the contaminant plume inside the area. The localization of the contamination was also promoted by the decreased level of groundwater due to the continuous pumping. The total pumping rates of the 3 wells was 300-320 m^3 per day, These wells were operated continuously for one year except during the winter period due to frost danger. The concentration of contaminants in the reinfiltrated water was less than 0.1 ppm. As can be seen in Figure 4, the operation time was not long enough to reduce the aquifer contamination to acceptable levels. However, by the beginning of September when the pumping was terminated, 56,500 m^3 of groundwater were treated (i.e., 1,400-1,500 kg of gasoline).

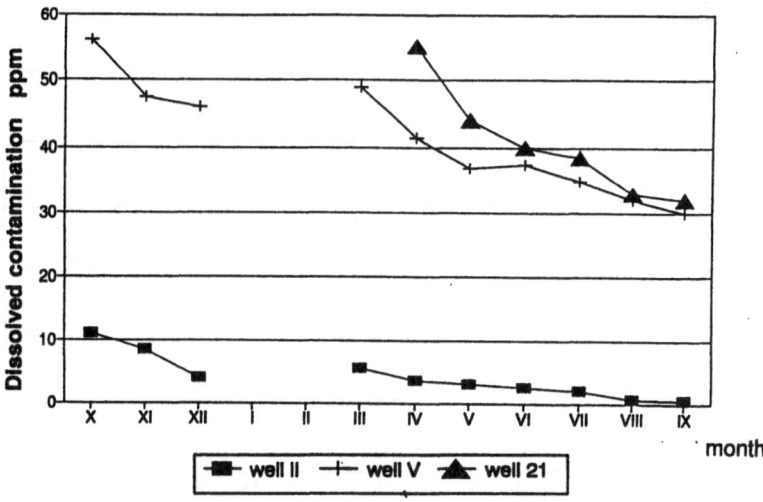

Figure 4. Dissolved gasoline concentration over time.

It was estimated to take about three years to clean up the gasoline contaminated area, including the source of soil contamination. A second phase of remedial operation was planned. A total of 15 wells, including the air extraction wells, were constructed that could alternatively be used as water extraction wells.

Venting for in-situ remediation of gasoline contaminated soil

Twelve extraction wells were installed in the vicinity of the eight gasoline tanks where the surrounding soil was heavily contaminated. Five of the wells were operated during the remedial activity.

The diameter of the extraction wells was 160 mm with screened segments that began at 3 meters below the surface and continued to the confining layer. Each of the wells had air removed by a side channel blower. The average parameter values were, vacuum: 0.11 bar and soil vapor volume: 190 m^3 per hour. The radius of influence for the wells in the soil texture present in this area is 30-40 meters, according to data in the literature. During the time of remedial operation about 80 metric tons of gasoline were removed. Figure 5 shows the volatile organic compound contents of the air extracted from the remedial wells. The site cleanup can not be considered accomplished. The residual gasoline in the soil continues to be a significant source of groundwater contamination.

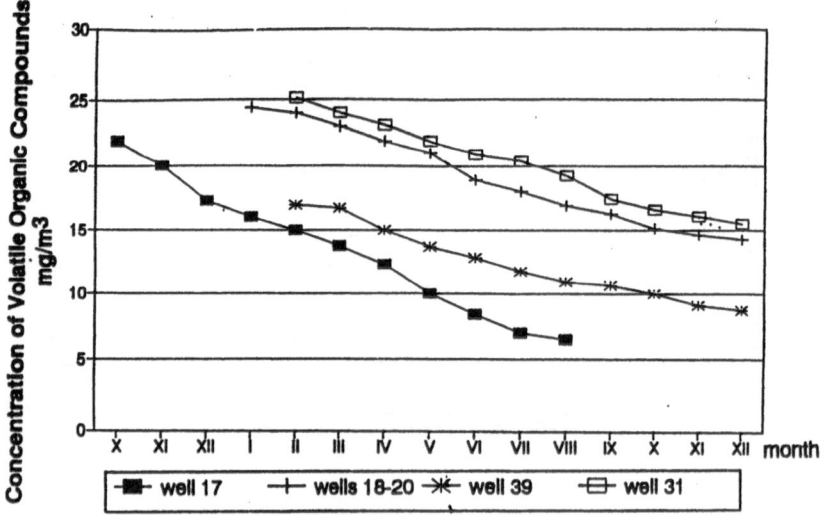

Figure 5. Quality of extracted soil gas over time.

Summary and Conclusions

The decontamination of soil contaminated with non-volatile compounds and which influenced a relatively smaller depth and volume was accomplished by the application of biological processes. At the gasoline polluted area, only the containment of the contamination was achieved. Regrettably, due to the lack of financial resources, the second phase of remedial action has not yet taken place. An immediate resumption of work would be necessary to more completely clean the area. The decrease of the contamination to an acceptable level will require the operation of two stripping towers with 4-6 water extraction wells and with soil aeration at 8-10 points. The efficiency of the technology could be enhanced by combining the vapor extraction with an air sparging technique, thereby providing treatment to both the saturated soil and groundwater.

References

Dott, W. (1992) Preinvestigations of biological decontamination potential assessment. Preprints International Symposium, Soil decontamination using biological process, Karlsruhe, Germany.

Jong S.C. (1991) Forced air ventilation for remediation of unsaturated soils contaminated by VOC. EPA/600/S2-91/016. U.S. Environmental Protection Agency.

Pitchford, A.M., Mazzella, A.T., Searbrough, K.R. (1988) Soil-gas and geophysical techniques for detection of subsurface organic contamination. EPA/600/S4-88/019. U.S. Environmental Protection Agency.

INNOVATIVE REMEDIATION TECHNOLOGY FOR

CONTAMINATED MILITARY SITES: A CANADIAN PERSPECTIVE

Igor J. Marvan
Bioremediation Group
Grace Dearborn Inc.
3451 Erindale Station Road
Mississauga, Ontario L5A 3T5
Canada

Abstract

A novel and innovative technology has been developed in Canada to clean-up soil contaminated with a wide range of organic pollutants, including aliphatic and aromatic hydrocarbons, chlorinated phenols, phthalates and pesticides. The technology, known as Daramend bioremediation, is based on the principle of favorably altering the environment within the soil matrix to enable microorganisms to degrade the pollutants. The approach consists of addition of solid-phase, biodegradable organic amendments prepared to have soil specific properties, including particle size distribution, nutrient profile and nutrient-release kinetics as well as careful control of process conditions, such as availability of moisture and oxygen. The full-scale application of Daramend has resulted in degradation of over 99% of petroleum hydrocarbons and chlorinated phenols and over 95% of polynuclear aromatic hydrocarbons (PAHs). The toxicity of soil as measured by earthworm mortality was completely eliminated.

Introduction

The efficient and economical clean-up of soils contaminated with petroleum hydrocarbons and other organic pollutants has become one of the most urgent environmental issues in Canada. There are large tracts of contaminated land at both active and closed industrial plants, railways and military installations. The contamination may often be linked to past material handling and waste disposal practices. Landowners are seeking cost-effective and environmentally acceptable alternatives to traditional soil remediation options such as incineration and landfilling.

NATO ASI Series, Partnership Sub-Series, 2. Environment – Vol. 1
Clean-up of Former Soviet Military Installations
Edited by R. C. Herndon et al.
© Springer-Verlag Berlin Heidelberg 1995

Up to the recent past, the most common approaches to addressing the issue of contaminated soil were excavation followed by containment of soil at landfills, and incineration. Landfilling of the soil obviously does not restore the soil in any way, but merely reduces the potential for contaminants in the soil to adversely affect other aspects of the environment, such as groundwater. Incineration, while effective in destroying the contaminants, at the same time, destroys all organic components of the soil and produces large quantities of ash with characteristics quite different from the original soil. Thermal processes such as incineration are relatively expensive ($300-$1,000/metric ton) and there are fears of releasing products of incomplete combustion to the atmosphere with the process off-gases. These considerations led to the prohibition of incineration for treatment of soils in some jurisdictions (i.e., Ontario).

In the past, bioremediation has been used with variable results and has therefore acquired a reputation as unpredictable and difficult to control. In general, application of the technology relied on the control of only one or a few process variables. The classical examples of bioremediation were addition of nutrients, addition of microorganisms and/or addition of oxygen (soil venting), or, in the best case, a combination of some of these variables. The variable performance of these classical approaches to bioremediation may be attributed to a lack of understanding and control of all of the complex relationships that exist between the bioremediation process variables in the real world.

Daramend Bioremediation

The key to consistently achieving effective bioremediation with the Daramend technology is an understanding of the complex relationships that exist among the soil matrix, contaminants, and microorganisms degrading the contaminants. A fundamental point often neglected in soil remediation is that all the biological processes take place in water held against gravity within soil pores. Hydrocarbon contamination will often coat part or all of a soil agglomerate with oil, thereby rendering the soil hydrophobic. Water holding capacities of contaminated and uncontaminated soil from the same site are often dramatically different. The hydrocarbons close the soil pores and coat nutrient-providing soil organic matter, limiting the supply of nutrients, oxygen and biologically available water, thus reducing the microbiological activity within the soil matrix. As a result, even in soil containing microorganisms with an enzymatic capacity to degrade the target compound, the rate of bioremediation may be too slow to be useful.

Application of Daramend bioremediation is based upon homogenous incorporation of organic amendments. The amendments improve the ability of the contaminated matrix to supply biologically available water, nitrogen, phosphorous, micronutrients, and oxygen to bacteria and other microorganisms with the capability to degrade organic pollutants. Concurrently, the amendments reduce the acute toxicity of the soil's aqueous phase by transiently adsorbing pollutants, and providing surfaces for microbial adhesion and development of biofilms. The composition of Daramend organic amendments is soil-specific and based upon the results of a thorough physical/chemical characterization of the soil or waste to be treated (e.g., texture, moisture retention, C:N ratio, nutrient profile, target compound identity and concentrations).

Performance Data

The performance of Daramend bioremediation has been extensively tested and demonstrated in full-scale remediation. In the last two years, Daramend technology has been applied to over 50 contaminated soils with a wide variety of properties and contaminants. Table 1 summarizes selected results from application of the Daramend technology.

Data showing the biodegradation of diesel fuel are presented in Figure 1. Field performance data indicate that diesel fuel can be rapidly degraded to low concentrations by application of Daramend. Very low (<50 mg kg) residual concentrations of the petroleum hydrocarbons were rapidly achieved. Similarly, rapid biodegradation of PAHs, including the most refractory 4-6 ring isomers, was observed (Figures 1 and 2).

In contaminated soil from a crude oil processing facility, treatment with Daramend reduced the concentration of high molecular weight aliphatic hydrocarbons. Performance data indicate that even the heaviest fractions of the oil (e.g., C_{35}-C_{44} aliphatic hydrocarbons) were rapidly degraded in soil during Daramend treatment (Figure 3).

Toxicity Removal

Regardless of the successful degradation of contaminants in soil as measured by analytical techniques, the question of whether a corresponding reduction of soil toxicity is achieved by bioremediation has been raised by environmental agencies in Canada. In order to provide the necessary answers, three separate bioassay approaches have been utilized: Microtox (Bulich, 1984), earthworm mortality and seed germination.

In order to cope with the high degree of variability between individual soils in terms of particle size distribution, particle composition, binding characteristics and moisture content, the principle of using a standard or reference soil was used in earthworm mortality and seed

Table 1. Daramend Performance Data

Parameter	Concentration (mg/kg) Initial [1]	Final [1,2]	Destruction Efficiency (%)	Treatment Time (days)
PAHs	1,442	35	97	240
PAHs	18,500	3,870	79	147
PAHs	1,488	33	98	207
4-6 Ring PAHs	1,273	32	97	207
chrysene	170	2	99	207
benzo(b)fluoranthene	140	5	96	207
benzo(a)pyrene	37	3.8	90	207
phenanthrene	117	n.d.	100	207
fluoranthene	410	2.9	>99	207
pyrene	338	3.1	99	207
PCP	680	4	>99	207
PCP	155	0.8	99	109
PCP	2,170	11	>99	280
Heavy Oil	2,372	22	>99	90
Diesel Fuel	8,700	35	>99	182
Hydrocarbons (C_{17}-C_{25})	1,003	198	80	138
Hydrocarbons (C_{26}-C_{34})	1,519	119	92	138
Hydrocarbons (C_{35}-C_{44})	853	107	87	138
Phthalates	4,350	26	99	130
DOP [3]	2,700	19	99	130
DNOP [4]	1,640	7.1	>99	130
p,p-DDT	684	1.9	>99	147

[1] Rounded to the nearest: 0.1 for concentrations > 0 and < 10

1.0 for concentrations > 10 and < 1000

10 for concentrations > 1000

[2] These values do not necessarily represent final residual levels; however, they provide an indication of remediation rates.

[3] bis-(2-ethylhexyl)phthalate

[4] di-n-octylphthalate

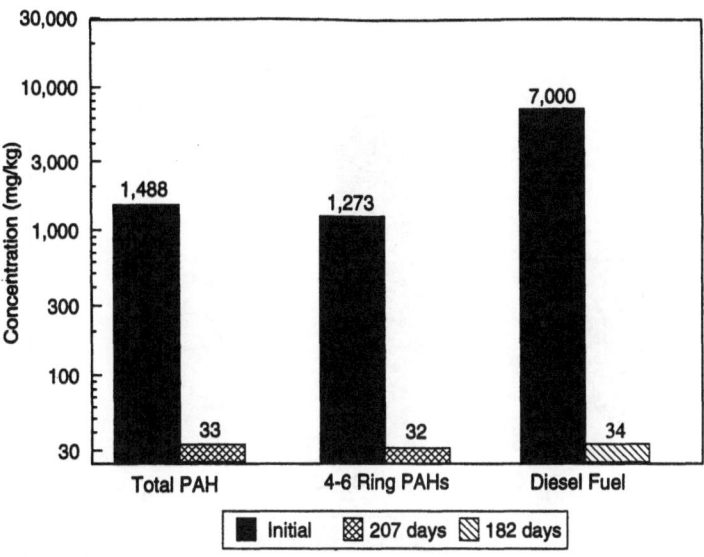

Figure 1. Biodegradation of PAHs and diesel fuel during Daramend bioremediation of industrial soil.

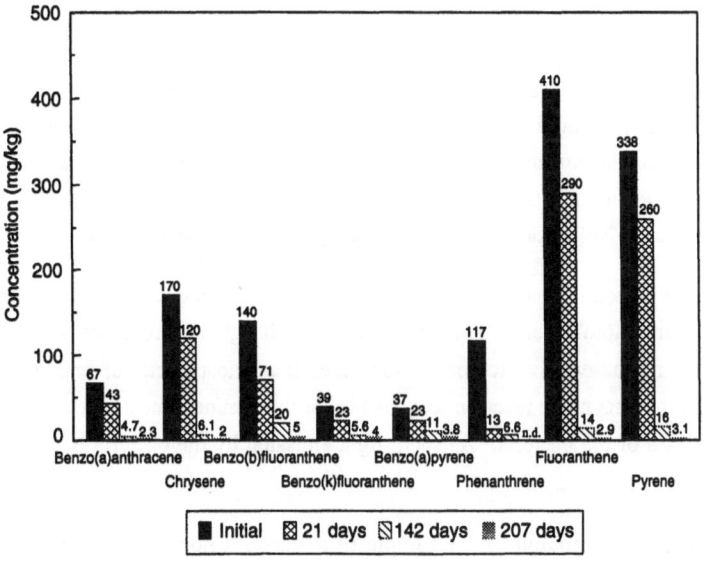

Figure 2. Biodegradation of high molecular weight PAHs in industrial soil during Daramend bioremediation

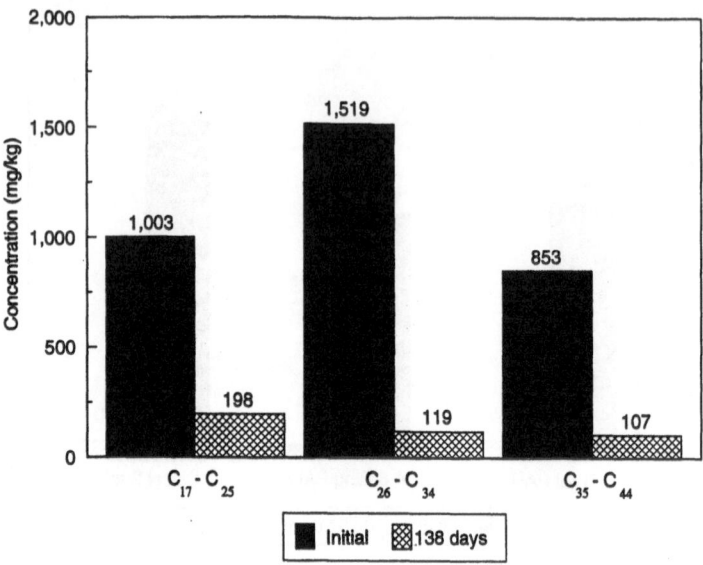

Figure 3. Biodegradation of petroleum hydrocarbons in clay soil contaminated by oil processing.

germination bioassays. The Microtox bioassay uses a standard liquid medium for the same purpose. It is important to note that the value of the toxicity test data result is not an absolute measurement of the toxicity itself, rather it is a measurement of the relative difference in magnitude between the toxicity before and after remediation. The use of standard soil ensures that the measured "soil toxicity" is not affected by factors unrelated to the soil itself.

Microtox is a bacterial bioassay that measures the toxic effect of a sample on a population of bioluminescent photobacteria. The photobacteria emit light as a result of respiration. These bacteria are more sensitive to toxic substances than most bacterial populations. A co-substrate in the enzymatic reaction responsible for the luminescence is linked to the electron transport chain of the bacteria. As such, light emission by the bacteria is directly linked to toxicity.

The effective concentration of the sample in a standard liquid media that causes a 50% reduction in metabolic activity, as measured by light emission (EC_{50}), is reported. The measured percent reduction in light emission is relative to the light emission observed in the control (standard liquid media alone). A higher EC_{50} indicates a less toxic substance.

Microtox toxicity was performed on soil (solid phase), and soil leachate. Microtox toxicity was reduced by a factor of approximately 100 in the bioremediated soil leachate, compared to the untreated soil leachate (Table 2). Microtox toxicity of the bioremediated soil was reduced by a factor of approximately 40 in comparison to the untreated soil (Table 3).

Table 2.   Influence of soil bioremediation technology on Microtox toxicity of soil leachate.

Sample	Mean 5 Minute EC_{50} (%)	Mean 15 Minute EC_{50} (%)
Untreated soil	0.41	0.27
Bioremediated soil	44	40

Table 3.   Influence of soil bioremediation technology on Microtox toxicity of soil.

Sample	Mean EC_{50} (%)
Untreated soil	0.25
Bioremediated soil	10

The earthworm mortality bioassay was conducted according to the OECD (1990) test method. It involved exposing common Night Crawlers or Dew Worms (*L. terrestris*), which are prevalent in the Canadian environment, to soil and monitoring their survival over time. Exposure was extended to a period of 28 days, as it is the prevailing opinion of the scientific community that the recommended period of 14 days is inadequate. An uncontaminated agricultural soil (Brookston series) was used as a control.

Results (Figure 4) indicated complete survival of earthworms in the bioremediated soil. No mortality (0%), or obvious signs of stress were visible after 7, 14, 21 or 28 days of exposure. Complete earthworm mortality (100%) was observed within four days of exposure to the untreated soil. There was no significant difference found between the bioremediated soil and the control soil with respect to earthworm mortality.

The seed germination bioassay was performed according to Green et al. (1989). It was conducted using species that are representative of the major market crops grown in Canada. Germination is monitored over a period of five days. The species used were radish (*Raphanus sativus*), oats (*Avena sativa*) and corn (*Zea mays*). A standard soil (Brookston series) was used as a control.

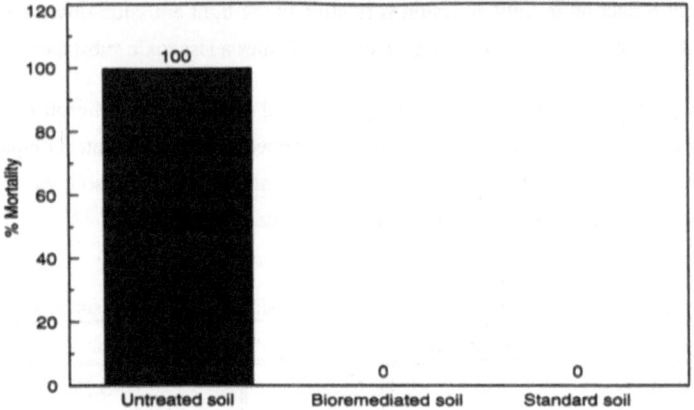

Figure 4. Influence of soil bioremediation technology on earthworm mortality

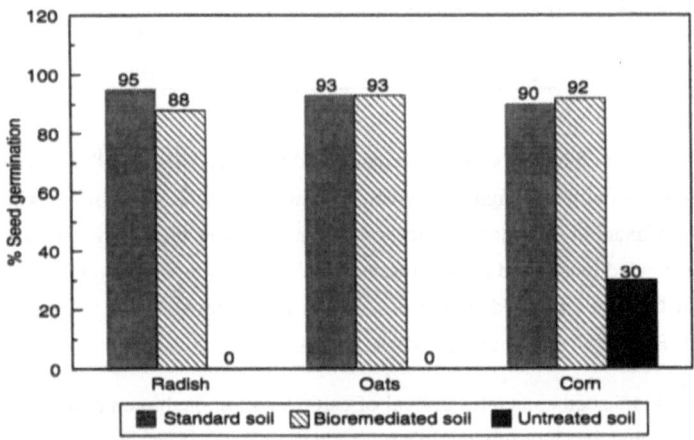

Figure 5. Influence of soil bioremediation technology on seed germination.

Results (Figure 5) indicated 88%, 93%, and 92% seed germination of radish, oats, and corn, respectively, in the bioremediated soil after five days of exposure. No germination (0%) of radish or oats, and only 30% germination of corn was observed in the untreated soil after five days of exposure. There was no significant difference between the bioremediated soil and the standard soil with respect to the germination of radish, corn, and oats. Such toxicity testing allows the effectiveness of bioremediation technologies to be more completely evaluated.

Bioremediation of soils contaminated with toxic organic compounds, with elimination of soil toxicity as measured by toxicity bioassays, ensures that the remediated soil will not adversely impact the ecosystem and provides a cost-effective solution to many site remediation problems.

Conclusion

Daramend bioremediation represents a major advancement in treatment of soils contaminated with organic pollutants in terms of the reduction in chemical concentration of target pollutants as well as the reduction/removal of toxicity from soil. The improved performance is achieved by studying and understanding site specific dynamic interrelationships among soil properties, contaminants and microorganisms and by controlling all the variables involved in the process.

References

Bulich, A.A. (1984) Microtox - a bacterial toxicity test with several environmental applications. In: *Toxicity Screening Procedures Using Bacterial Systems*. D. Liu and B.J. Dutka (Eds.). Marcel Dekker, New York. p.p. 55-64.

Green, J.C., Bartels, C.L., Warren-Hicks, W.J., Parkhurst, B.R., Linder, G.L., Peterson, S.A. and Miller, W.E. (1989) *Protocols for short-term toxicity screening of hazardous waste sites.* EPA 3-88-029.

Organization for Economic Co-operation and Development (OECD) (1990) Guidelines for testing of chemicals: 207E, *Acute Toxicity Tests*, (1984). ISBN 92-64-12221-4.

Seech, A.G., Marvan, I.J. and Trevors, J.T. (1993) On-Site/Ex-Situ Bioremediation of Industrial Soils Containing Chlorinated Phenols and Polycyclic Aromatic Hydrocarbons. *In Situ and On-Site Bioremediation: An International Symposium*, Battelle Institute, San Diego, California.

BIODEGRADATION OF PETROLEUM HYDROCARBONS AFTER

THE DEPARTURE OF THE SOVIET ARMY

Frantisek Kastánek
Institute of Chemical Process Fundamentals
Czech Academy of Sciences
165 02 Prague 6
Czech Republic

Katerina Demnerová
Prague Institute of Technical Chemistry
Institute of Biochemistry and Microbiology
160 00 Prague 6
Czech Republic

Summary

The contamination of soils and groundwater caused by the operation of army vehicles by the former Soviet Army (SA) can be considered a typical example of wide-spread, multiple point-source pollution. It has been shown that mixed populations of autochthonous microbes are particularly suited for the remediation of soil and groundwater. In fact, petroleum contaminant levels in groundwater were found to have been reduced to permissible levels within approximately 6 months, though it is necessary to continue monitoring contaminant levels after treatment. Soils are treatable in this fashion as well. Concentration of contaminants have been reduced from tens of thousands of mg kg^{-1} to approximately 250 mg kg^{-1} within a three month summer season in favorable soils (i.e., sand, sandy loam, metamorphosed clays, gravel).

Introduction

After the fall of the totalitarian regime in Czechoslovakia in 1989, the SA gradually departed from sites they had occupied since the Soviet invasion of 1968. The departure of the armies from the Czech Republic was completed in June of 1991, but they left behind over 80 sites mainly used for barracks, airports, military field areas or strategic weapons stockpiles. For more than 20 years, these areas were contaminated with toxic and hazardous substances,

NATO ASI Series, Partnership Sub-Series, 2. Environment – Vol. 1
Clean-up of Former Soviet Military Installations
Edited by R. C. Herndon et al.
© Springer-Verlag Berlin Heidelberg 1995

mainly oil products, due to the complete lack of care in handling these materials. Releases primarily resulted from heavy machine maintenance, fueling procedures and leaky underground storage tanks. The cleanup efforts at one such location in the town of Vysoké Myto in Eastern Bohemia is described in what follows.

Characteristics of the Site

Monitoring and hydrogeological investigations were carried out before the beginning remediation of these areas. The main conclusions resulting from this work were as follows:

1. Storage facilities for petroleum substances leaked into soil and groundwater.
2. Soils in this area were found to be contaminated to an average depth of 1 m^2 over a total area of 22,000 m^2. The concentration of contaminants ranged from 350-59,620 mg kg^{-1}.
3. Petroleum substances from the aeration zone had penetrated the shallow upper Turonian watery layers and contaminated them over an area of more than 2 km^2. At the centers of pollution the concentration of petroleum substances ranged between tens and hundreds of ug L^{-1}.
4. The mobility of the petroleum substances depends on several factors: a) the quality of the substance (in the Vysoké Myto region it is primarily light petrol), b) the character of the surrounding soils (in this locality the soils are not very porous), and c) the amount of water contained within the soils, along with the degree of saturation of the petroleum substances. Due to the fact that ground composition is very heterogeneous (Ca-clay originating from the disintegration of marl, with pockets of sand), the movement of the contaminants is also extremely heterogeneous with respect to both time and space. The petroleum substances did not reach the groundwater at once over the whole area, but rather they accumulated in the aeration zone and were gradually washed out by rain soaking into shallow, water-bearing zones. Contamination originated from point sources in the immediate vicinity of storage tanks and filling stations. The actual spread of the petroleum substances in the watery environment is caused by 1) the flow of the water, 2) the convective dispersion caused by the density difference of the liquids, and 3) molecular dispersion. According to hydrodynamic calculations, the spread of the contaminants in the watery environment can be considered to be very slow, and local leaks of petroleum substances into the surrounding ground form relatively isolated centers of pollution.

Over the whole cleanup area the subsoil consists of a uniform layer down to about 0.8 m of clay, gravel, stones, and, in places ground made up of the remains of concrete and asphalt surfaces. Strata vary according to individual drill holes between 0.8-9.0 m; there is loess, loamy clay (zones I and II) and sometimes compact calcareous loam and/or compact claystone (zone III). The groundwater level is at an average depth of 5.5 meters.

Concentration of Petroleum Substances

Concentration of petroleum substances in the liquid phase in samples taken from drill holes at the groundwater level varied between 0.08 -16.0 mg L^{-1}, and between 0.02-40.8 mg L^{-1} at a depth of 12 m, expressed as total non-polar hydrocarbons. Concentrations in soil samples in surface zones to the depth of 1 m varied between 350-59,620 mg kg^{-1} of dry soil. It was obvious from the screening of water and soil samples that the concentration of the contamination was locally heterogeneous, and that it was intrinsically related to the previous locations of the sources (e.g., fuel filling stations, storage tanks).

Selected Remediation Methods for Petroleum Products

Biodegradation was chosen as the method of remediating the petroleum substances which were primarily composed of non-polar hydrocarbons. Three methods were compared and previously tested in a pilot experiment at this location (experimental field of area 10x10 m):
- use of selected cultures of *Pseudomona putida* (commercial product);
- initiation of microbial activity by stimulating growth of microorganisms (already present in contaminated soils) with nutrients and oxygen; and
- use of autochthonous microbial mixed populations.

We choose the last of these methods as the most suitable for the given location. The actual technology of the cleanup consists of:
- surface decontamination of the soils *in situ*;
- decontamination of the soils at the decontamination area; and
- sanitary pumping of contaminated water and the use of bioreactors.

Surface Decontamination of the Soils *in Situ*

Fifteen regions were defined where the surface zone of the mineral profile had been severely polluted. They were typically located in the garage and parking areas, the sites of mobile stores of fuel and lubricating oil, and the surrounding areas of the wash line. The total area covers 22,000 m^2. In these locations, the upper part of the ground profile is loosened by machine to a depth of 1.0 m and decontamination of the soil is carried out with the application of bacterial and nutrient solutions (the treated soil was periodically loosened for the purpose of aeration, and occasionally sprayed with solutions of nutrients and soil amendments).

Decontamination of the Soils at the Decontamination Area

At 17 sites it was necessary to extract the soils and to decontaminate them in a safe decontamination area. The reason for this method is the unfavorable hydrophysical characteristics of the ground under the bottom of the former underground storage tanks (compressed loess and loamy clay), which made effective decontamination by any of the available *in situ* methods impossible. The earth was extracted to a depth determined by the state of the oil pollution at each site. Altogether, 9,000 m^3 of the earth were extracted and taken to a water- and oil-proof decontamination area of about 10,000 m^2, built on-site. The area has a down-gradient leading to a sump. Here there was a built-in device for the collection of the free phase of the oil products and the pumping of the filtrate back to the decontamination area. The layers of processed earth were about 0.5 m high, and the decontamination was carried out with the application of biodegradation and nutrient solutions. The soil in the layer was periodically loosened for the purpose of aeration. Two decontamination cycles of 3 months each were used; in each cycle 4,500 m^3 of earth were processed.

The threshold level of 500 mg kg^{-1} of non-polar hydrocarbons was considered as excessive pollution, in need of treatment using our surface decontamination and extraction method; soil polluted below this concentration was not decontaminated.

Sanitary Pumping of Contaminated Water and the Use of Bioreactors

It was decided to focus cleanup work only in the most polluted areas (above the level of 1 mg L^{-1} of the concentration of non-polar hydrocarbons). It would be technically and economically unrealistic to attempt the complete remediation of oil pollution in the aeration zone and the groundwater. Due to the potential threat of pollution of the deeper water zones and the need to determine the effectiveness of the proposed sanitation work, a long-term system of operations was set up to monitor the quality of the water in the whole affected region for a period of 5 years.

The extraction of contaminated water varied in the range of 0.05 - 0.1 L sec^{-1}. The water was pumped to pass through the bioreactor and was sprayed back or conveyed into neighboring drill holes.

Laboratory preparation and application of the biodegradation solution in situ: The basis of this method is the use of autochthonous microorganisms. The isolated microorganisms were arranged taxonomically, and species not considered pathogenic or conditionally pathogenic were selected and grown in the laboratory. The survival of these microorganisms in soil and water was determined.

The resultant preparation for the Vysoké Myto site consisted of three bacterial cultures capable of degrading oil hydrocarbons: DBM 155 genus Acinetobacter, DBM 163 genus Acinetobacter, DBM genus Micrococcus. A mixture was prepared in the ratio of 1:1:2 respectively.

Efficiency of degradation of non-polar hydrocarbons (light petrol) with this mixed culture was determined under pilot plant conditions with a sample of native soil (calcareous loam, zone III) containing an initial concentration of 720 mg kg^{-1} of non-polar hydrocarbons. Degradation was carried out at different temperature levels (4°C, 18°C and 28°C). After 42 days of treatment, the degradation efficiency was 53%, 75% and 82%, respectively.

Application at field conditions, soil decontamination: 50 L of the inoculum containing about 10^9 cells per ml was placed in a cistern with 2.5 m^3 of service water. The cistern was aerated with compressed air, which ensured that the contents were thoroughly mixed and oxygenated during transport from the production location to the cleanup area. Industrially produced fertilizers were added as a source of nitrogen and phosphorus.

Application at field conditions, water decontamination: 50 L of the inoculum were poured into 2 m^3 of water in the bioreactor and aerated either with compressed air, or through the use of an ejector. The liquid was pumped and recycled over the ejector, and air was introduced under pressure. The dispersion of bubbles in the reactor created a high mass transfer area.

Bioreactor: The bioreactor was an open vessel with a capacity of 2 m^3 with built-in netting made of perforated polyamide. The cells were spontaneously immobilized on this netting. Gradually the spontaneous immobilization of the bacteria took place in the interior of the bioreactor. The total growth of biomass over a period of some 2 months was approximately ten-fold over the original quantity of bacteria introduced. Water was continuously pumped from drill holes into the bioreactor through a vessel heated using a solar energy panel. Nutrients in solid phase (ammonium phosphate) were also periodically added. After passing through the bioreactor, water returned to the infiltration drill holes. A series of bioreactors were applied. The residence time in bioreactors was approximately 25 hours. The content of nutrients (phosphorus) and dissolved oxygen in the bioreactor was controlled, and input and outlet values of non-polar hydrocarbons was monitored.

Results and Discussion

It has been shown that this method is particularly suitable for cleaning contaminated groundwater; the water taken from drill holes was cleansed to a contaminant concentration of less than 0.05 mg L^{-1} in a period of 6 months. However, it can be expected that over a period of time the content of oil substances in these drill holes will rise again, due to the migration of underground water.

The extracted soil was taken to the operating area, which was divided into 16 sub-fields to monitor the time history of decontamination. Before decontamination, each field contained different compositions of oil hydrocarbons, as the soils were brought from different places. These fields were evaluated in the same manner at the end of the decontamination. Table 1 reports the concentrations of non-polar hydrocarbons in the extracted soils treated through biodegradation at the biodegradation area.

It is clear from the results that the biodegradation method can be considered an efficient method of decontamination of soils of favorable material composition and structural characteristics (Zone I and Zone II), and that bioremediation continued even during winter months. The relatively poor results obtained with soils excavated from Zone III may be due to the relatively non-porous soils in that zone.

Table 1. Concentrations of non-polar hydrocarbons in soils.

Zone	Initial Concentration, mg/kg		Final Concentration, mg/kg	
	mixed sample	maximum	mixed sample	maximum
I	2388	9116	101	742
II	2825	12456	237	2163
III	438	1490	213	1742++

Time of treatment - 6 months (October-March, no wetting was done in January and February)

++   Discrepancy in the maximum concentrations before and after the application of biodegradation could be due to erroneous sampling and/or low efficiency of bioremediation applied on samples of compact calcareous loam.

Bioremediation of polychlorinated biphenyls (PCB) in groundwater: During continuous monitoring on-site, the waste accumulation of PCBs in pumped waters was recently found in one of the newly opened monitoring drill holes situated near the border of a decontaminated area. The PCBs found are primarily composed of those with 3 and 6 chlorine atoms per molecule. The source of the PCBs is not known, but could, perhaps, be attributed to leakage from a capacitor storage facility; these capacitors were filled with PCB liquids. The storage facility was located close to the military area. Investigations of the contaminated area revealed that the contamination is spatially limited, the highest concentration of which is situated at the boundary between the layer of sandy loams and the non-penetrable claystone base layer. However, the PCB contamination has reached the level of the groundwater. Contaminated soils have been excavated and are provisionally stored at a safe deposit area *in situ*. The concentration of PCBs in pumped waters ranges from 3,000-12,000 ng L^{-1}.

Method of decontamination: Groundwater from the drill hole was pumped to the decontamination units. Based on previous laboratory work, the decision was made to apply the method of decontamination based on sorption of PCBs on selected sorbents, with subsequent bioremediation of the sorbed PCBs. Decontamination units were designed to serve as two independent units for:

a) water with high content of natural solid earth particles (above 0.1%); and
b) "pure" water.

In the first case, the contaminated water was pumped to the sorption vessel where the sorption phase took place. Activated bentonite (activated with Fe^{3+}) at pH 3 was used as a sorbent. When the sorption process was finished, the slurry was neutralized and the proper flocculent was added. It was then discharged to a series of sedimentation vessels and "pure" water flowed through the final filtration station (activated carbon), and was finally discharged into the environment. The unit operated semi-continuously. The capacity of the unit was 10 m^3 per day. The concentration of PCBs of the decontaminated water did not exceed 200 ng L^{-1}.

In the second case, the contaminated water flowed directly through a series of adsorption columns filled with activated carbon. The outlet concentrations of PCBs also did not exceed 200 ng L^{-1} (Czech limit for possible introduction of waters to the environment). The capacity of that unit was 20 m^3 per day.

Biodegradation: The waste slurry (after the sedimentation of bentonite and/or the content of waste activated carbon) containing PCBs in the range of 8-30 mg kg^{-1} of dry material was treated in a series of bioreactors at strictly aerobic conditions. The bioreactors were open vessels (volume approximately 2 m^3) which were aerated with compressed air. The volume of solid phase in a slurry did not exceed 10%.

For biodegradation a proprietary bacterial strain was used. In about 40 days the initial content of sorbed PCB decreased to approximately 2 mg kg^{-1}, the concentration at which the slurry could be deposited with the contaminated soils stored *in situ*.

An important result from the laboratory research was that at strictly controlled laboratory conditions (i.e., temperature, pH, intensity of aeration, sterile conditions), PCBs with a low chlorine content are degraded, while forms with 6 or more chlorine atoms remained almost intact. However, at field conditions the forms with 6 or more chlorine atoms were degraded with high efficiency as well. This phenomenon possibly can be attributed to the complementary activity of "wild" strains of microorganisms proliferating at field conditions.

Preparation of biodegradation microorganisms: Eluates from contaminated soils containing soil microorganisms were used as an inoculum for a mineral medium which contained pure biphenyl as the sole carbon source. Of the 30 strains collected from the primary mixture of microorganisms, 5 of them survived the biphenyl selection and were tested for their ability to biodegrade the industrially produced PCB mixtures. The two strains possessing the highest PCB degradation activity for forms with 3 chlorine atoms per molecule were selected from this group. Both strains were classified among Pseudomonas.

ENVIRONMENTAL PROBLEMS AT FORMER SOVIET MILITARY

INSTALLATIONS IN THE CZECH REPUBLIC

Ivan Landa, Oldrich Mazac and David Redlin
ECOLAND
P.O. Box 512
111 21 Prague
Czech Republic

1. Introduction

Although the Soviet Army (SA) withdrew from (former) Czechoslovakia in 1991, some significant environmental damage will remain for a number of years. In fact, some of the damage may never be remediated. Recent experience with site remediation following the Soviet occupation (1968-1991) of the Czech Republic has led to the following:

a) an understanding of contaminant transport in a rock medium;

b) the development and use of methodologies for risk analysis;

c) the use and evaluation of various remediation methods;

d) the development of criteria for optimum protective measures; and

e) the application of new survey methods, mathematical modeling techniques, and various site monitoring approaches.

This paper will help not only to describe the consulting and remediation work of our company, as well as other Czech companies, but also to analyze the costs of various site characterization and remediation activities.

2. Historical Aspects, Distribution of Soviet Military Sites and Environmental Damages

The August 1968 invasion by the SA had a negative impact not only on the political, economic and military situations of our country, but also on the environment. The SA made use of former bases of the Czechoslovak Army (e.g., Vysoke Myto, Mlada Boleslav, Hradcany, Milovice), and built new garrisons, or training zones, where family members of the Soviet soldiers were accommodated (e.g., Milovice and Libava area). These areas were used to provide for the needs of the military families, including schools, theaters, laundries, etc.

NATO ASI Series, Partnership Sub-Series, 2. Environment – Vol. 1
Clean-up of Former Soviet Military Installations
Edited by R. C. Herndon et al.
© Springer-Verlag Berlin Heidelberg 1995

A majority (76%) of the Soviet military installations was used by the Czechoslovak Army before the Soviet occupation in 1968, while the rest (24%) was built by the Soviets. About 71% of the former Czechoslovak Army sites were completely occupied by the SA; the rest were jointly used by the Czechoslovak and Soviet Armies with the SA participation varying between 5% and 95%.

The distribution of the SA sites is shown in Figure 1. With respect to water-bearing characteristics and hydrogeological conditions, about 60% of the SA sites were, unfortunately, located in areas underlain, to a great extent, by porous and karst formations with significant discharge of 5-100 liters sec^{-1}; 25% were in areas underlain by fissure formations with the mean discharge up to 1 liter sec^{-1}; and 15% were in areas underlain by fissure-porous formations with the mean discharge of about 5 liters sec^{-1}. Through an agreement between the Czechoslovak and Soviet governments, the SA started to leave the Czech Republic in March of 1990, and completely withdrew from the Czech Republic by August of 1991. Recent estimates indicate that about 74% of the former SA sites are owned by the private sector, 19% by the Czech Army, and the remainder by joint civil and military owners. During and after the SA departure, a survey was undertaken, and the total environmental damages were estimated at over $170 million (USD), with specific damages caused to groundwater and rock structures placed at about $66 million (USD). Initial remediation action began prior to completion of the survey.

It is obvious that the pollution sources have adversely affected the environment for most of the 23-year occupation, though, in many cases, pollution plumes have had a stationary character. In less favorable conditions, however, the plumes have infiltrated to great depths and distances.

3. Main Causes of Pollution

The environmental impact of the SA occupation of the Czech Republic was substantial. The main pollution causes are described in what follows.

3.1. Technical Aspects

The individual base structures or their component parts were constructed mainly in the Soviet Union or other Eastern European countries according to Soviet Standards (the so-called GOST), which differ from Czech Standards (CSN). Most structures were constructed without regard to Czechoslovak building regulations. It is suspected that some buildings were not even constructed to suit Soviet civilian sector standards.

DISTRIBUTION OF THE SOVIET MILITARY SITES-CZECH REPUBLIC

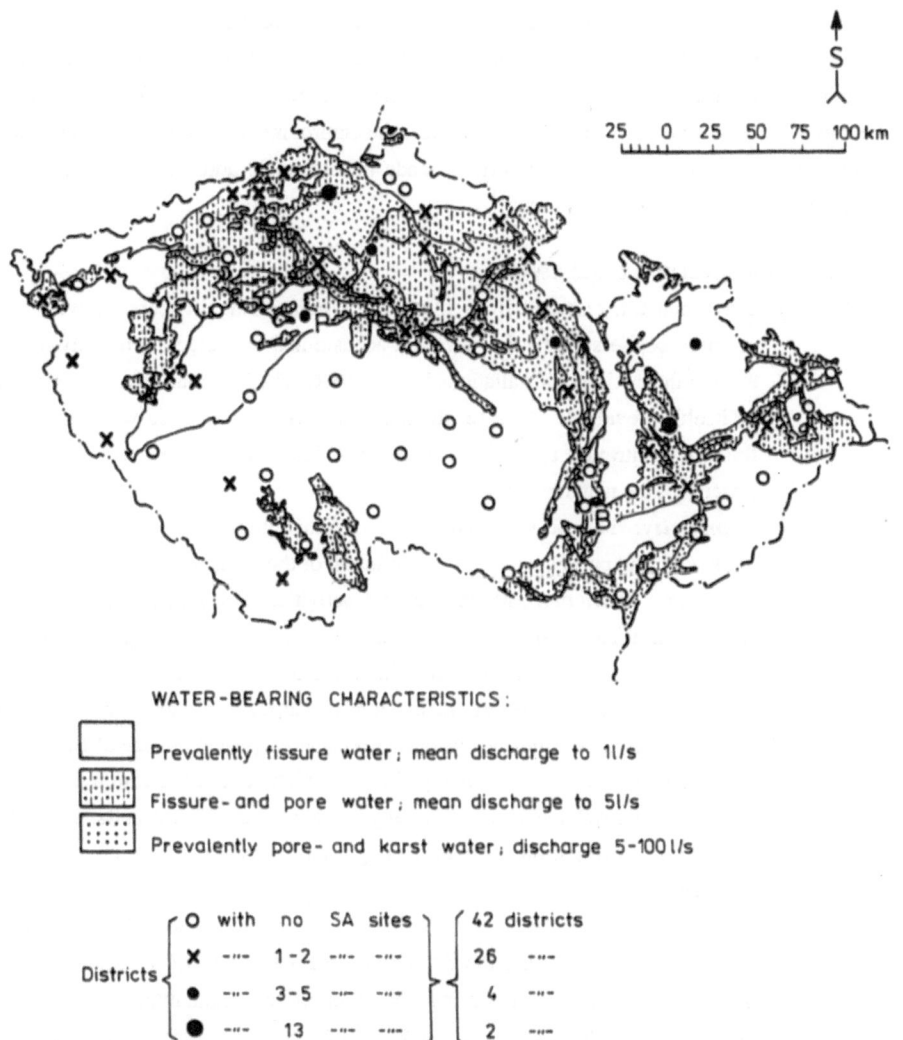

WATER-BEARING CHARACTERISTICS:

☐ Prevalently fissure water; mean discharge to 1l/s

▨ Fissure- and pore water; mean discharge to 5l/s

▨ Prevalently pore- and karst water; discharge 5-100 l/s

Districts
○	with	no	SA sites	42 districts	
✕	-"-	1-2	-"- -"-	26	-"-
●	-"-	3-5	-"- -"-	4	-"-
⬤	-"-	13	-"- -"-	2	-"-

ECOLAND PRAHA, 1994

Figure 1

3.2. Military Technological Aspects

The SA used á range of products and materials for the construction and maintenance of their sites which were not commonly used in the civilian sector, or which were internationally prohibited for use (e.g., DDT). Lax organizational requirements and record keeping allowed many materials, especially fuels, to contaminate the environment in areas where the materials were used or stored. The SA authorities made no efforts to address the procedural and technological causes of the pollution.

3.3. Psychological and Legal Aspects

a) *Environmental culture:* Both the chief commanding officers as well as the ordinary soldiers of the SA were not accustomed to protecting the environment, even in their original installations in the USSR. The idea that "nature can take care of itself" prevailed. As a result, toxic or harmful materials were disposed of through infiltration to the rock medium. Most soldiers did not understand the toxicity of the materials they were using.

b) *Anonymity:* This was a factor which contributed to the illegal disposal of pollutants, and consequently, to the pollution of the environment.

c) *Information restrictions:* Much of the SA activities were kept secret. The transfer, accumulation and analysis of input information about environmental conditions was, for the most part, concerned with the control of potable water for the SA. There are no data confirming the existence or absence of any monitoring system built by the SA. Monitoring systems subsequently placed outside of two SA sites detected significant groundwater pollution, which could have resulted in damage to the drinking water sources for Karany and Sojovice. The SA authorities refused to yield all information, or even to begin any remedial action concerning this problem.

d) *No disciplinary action:* Damage to the environment was not grounds for disciplinary action within the SA. Punishment for the loss of a pair boots was much greater than that for pouring a barrel of oil into a road ditch.

e) *Estrangement:* This manifested itself in the psychology of the SA as a feeling that they were in a foreign country which they would eventually leave. This was best illustrated during the SA departure in 1991 when soldiers freely poured fuel residuals onto the ground (e.g., in Vysoke Myto).

The SA did not create regulations for prevention of pollution nor for subsequent active remediation of the environment. These modes of behavior were quite common for SA soldiers.

4. Classification of Pollution Sources

The pollution sources at SA sites in the Czech Republic can be characterized according to the following: their geometry; the transport capabilities of pollutants; the hydrogeological characteristics of rocks and aquifers; the military activity causing the pollution; the impact on the geological environment; and the types of the pollutants.

4.1. Geometry of the Sources

The geometric attributes of the sources can be described as:

a)   point (e.g., fuel stores, workshops, dumps, septic systems);

b)   linear (e.g., canals, pipelines);

c)   areal (e.g., large-scale fuel stores, airport runways);

d)   macroareal or regional (e.g., pollution plumes around pumping stations, target areas at shooting-ranges, combat tank exercising grounds).

4.2. Transport Capabilities of Pollutants

The transport capabilities of pollutants may be:

a)   *high* in the case of fecal pollution, chemical weapons contamination, chlorinated hydrocarbons, and some kinds of radioactive wastes; or

b)   *low* if influenced by sorption and degradation processes (e.g., grease).

4.3. Hydrogeological Characteristics of Rocks and Aquifers

The subsurface characteristics at these sties varies and consists of:

a)   rocks characterized by *low permeability*, in which:

i - the pollution is spread through a system of fissures. The pollution may be concentrated and not widely detected over large areas. Pollution plumes are located in heterogeneous rock, and are, as a rule, local. Aquifers are located in crystalline rocks or in sedimentary rocks with fissured permeability;

ii - the pollution lies within a zone of aeration having low porous permeability. These pollution plumes are local. The zone is insulated by a sedimentary overburden. Remedial costs depend on the penetration depth of the pollution plume;

b) aquifers, areally limited, characterized by *medium-to-high permeability*, with relatively thin aeration zones. The pollution is shallow, concentrated, and relatively easy to remove. This usually occurs in Quaternary aquifers, often with local-to-regional significance for water management;

c) aquifers of a regional scale, characterized by *high permeability* with large aeration zones and/or having large overburden thickness. The pollution is deep and its removal is costly. These aquifers are of Cretaceous age and are of great importance for water management;

d) aquifers of variable size, characterized by *low-to-middle permeability* and significant overburden thickness. Pollution is deep, the aquifers have low permeability, and the pollution can typically only be removed over a long period of time. These deep aquifers are mainly of Tertiary age and usually have local importance for water management.

4.4. Military Activities Causing Pollution

The most significant sources of pollutants are listed below:

a) production of energy for military installations;

b) development, production and testing of weapons and weapons systems (e.g., at the military-industrial complex);

c) accommodations of soldiers and their families, administrative buildings, command posts and headquarters;

d) stores of military materials, including fuels;

e) construction activities and the production of building materials;

f) raw material mills, finishing plants, etc.;

g) transportation activities (shipping by boat, road, rail, and air);

h) troop and equipment combat exercising grounds and grounds for testing weapons; and

i) communication systems.

4.5. Impact on Geological Environment

The military activities (described in Section 4.4) allowed contaminants to enter the environment. While some of these contaminants were normally present at low concentrations, other contaminants have been detected at very high concentrations.

As the pollutants move through the geological environment, they can be:

a) eliminated due to natural degradation (e.g., biodegradation, oxidation, ultraviolet radiation, evaporation);

b)   dispersed, resulting in a decrease in pollutant concentration;

c)   accumulated without changing their properties (e.g., some organic compounds such as DDT);

d)   transformed into new physico-chemical compounds which may have different transport, toxicological, and hygienico-epidemiological properties.

These military activities had other adverse effects as well:

a)   man-made structures, such as waste dumps, communication banks, subsurface walls, rocket silos, tunnels, bunkers, etc., which have specific physico-mechanical, hydrogeological, and hydrochemical properties, often had direct, negative influences on the natural properties of an area; and

b)   the negative influence on the pre-occupation hydrogeological, hydrochemical, and hydrogeophysical conditions at SA installations can still be seen in verified changes in:

   i)   filtration parameters in deep excavations;

   ii)   infiltration and evaporation conditions at military airports;

   iii) groundwater quality in the vicinity of storage areas, waste dumps, motor pools, etc.; and

   iv)   groundwater infiltration from drainage systems, water pipes, rivers, etc.

4.6. Types of Pollutants

A wide variety of pollutants are commonly found at military sites. In addition to the products commonly used for industrial activities, the pollutants found at many of the Soviet installations had a distinctly military application. Pollutants included:

a)   chemical weapons materials;

b)   chemical weapons deactivation materials;

c)   products for the cleaning and maintenance of weapons;

d)   explosives;

e)   dichlorethane, monoethanolamine, tear gas (chlorpikrin);

f)   dichloramine, chlorinated lime, DDT, cooling liquid; and

g)   fuel for rockets.

5. Site Investigation Methods

The basic methods applied for solving hydroenvironmental problems at former Soviet military installations have been both hydrogeological (hydrodynamic and hydrochemical) and geophysical in approach.

Hydrogeological methods have been used mainly in analyzing the results of hydrogeological mapping and boreholes (i.e., hydrodynamic tests). Following the initial testing, some boreholes can be used for monitoring and/or remediation.

Applicability of the geophysical methods (i.e., airborne, surface, underground methods) depends mainly on the geophysical properties of the rock medium, groundwater, aquifers and pollutants.

The use of geophysical methods can:

a)   detect pollution sources and the extent of pollution in the zone of aeration in aquifers, or even in the air;

b)   determine or estimate the geological and hydrogeological conditions of the surveyed locality; and

c)   locate buried ammunition, metal objects, etc.

The most frequently applied geophysical methods in the Czech Republic have been geoelectrical methods (resistivity methods - vertical electrical sounding and profiling, electromagnetic, ground penetrating radar), magnetometric methods, seismic methods (refraction), and gasometric methods (using ECOPROBE, PORTAFID). Less frequently applied methods are, thermometric methods and gravity testing.

6. Site Remediation Technologies

The site technologies used (or planned for use) for the first remediation stage in the Czech Republic include: pumping (34%), monitoring (33%), biodegradation (30%), and, to a much lesser degree, thermodegradation (3%).

The following information summarizes the site technologies used for remediation of the Soviet installations in the Czech Republic. Provided is information on the principles of the technologies, companies which have used them or intend to use them, and the advantages and disadvantages of each technology. The remediation activity at the Soviet installations initiated the development and use of many new technologies in the Czech Republic. For example, in the case of biotechnologies, it was only the EKOL Praha company which had the necessary licenses for the practice of biodegradation at the start of the Czech remediation activities. EKOL began the remediation of the first locality (Frenstat) in 1991. By 1992, there were four companies using biodegradation (EKOL Praha, VUAB Roztoky, ALPHA BIO Praha, VUGI Brno). Today, biodegradation methods are being used by 15 companies (AQUA PLUS Praha, ENVISAN Ceske Budejovice, BIOASAN Praha, ECOHELP Brno, DEKONTA Kladno, et al.). A similar phenomena occurred in the case of venting (in 1992, just AQUATEST Praha, in 1993 KAP PRAHA and SAKOSTA Praha; by 1994, HYDROGEOLOGIE Chrudim, DEKONTA Kladno, APV Praha, et al.). An outline (A-G) of the applied remediation methods is presented including a description of the methods.

A. NATURAL DEGREDATION

B. HYDROGEOLOGICAL METHODS

    B.1. Hydrodynamic

    B.2. Venting

C. BIOTECHNOLOGY

    C.1. Stimulation of natural biodegradation processes

        C.1.a by artificial nutrients and other materials

        C.1.b by natural nutrients

        C.1.c by bioventing

    C.2. Application of allochthonous bacterial strains

    C.3. Application of autochthonous bacterial strains

D. COMBUSTION

E. SOLIDIFICATION

F. STORAGE/DISPOSAL

G. UNSPECIFIED METHODS

Each category of remediation method includes information on the following: (P - principle, E - economic/financial considerations, D - disadvantages, A - advantages, R - remarks)

NATURAL DEGREDATION

P: Degradation of the oil products (OP) is caused by natural microbial processes, venting, ultraviolet radiation, oxidation, etc. The rate of degradation depends on hydrogeological conditions, pollutant concentration, type of OP, and other factors. Increasing the rate of degradation is the goal of many remediation technologies.

E: The price is relatively low.

D: OP degradation may require decades. The processes are typically not directly controllable.

A: The processes are spontaneous.

R: Plant extraction (plant uptake) techniques (experiments have been performed in the Czech Republic with sunflowers, rushes, and other plants) for remediation of surface oil pollution can be included with this method.

HYDROGEOLOGICAL METHODS: Hydrodynamic (B.1)

P: The OP are pumped out either in phase, or dissolved together with polluted groundwaters.

E: This is the method most frequently used by AQUATEST Praha, GEOTEST Brno, UNIGEO Ostrava, KAP Praha, STAVOPROJEKT Pardubice, TOPGEO Praha, and others.

D: The method is both time consuming (e.g., at the former Soviet installation Hradcany, KAP Praha is planning a remediation effort that will last more than 10 years) and expensive (e.g., pumping on one borehole costs about $50-80 (USD) daily) even if OP release is initiated by vapor (EKOL Praha, ECOLAND Praha), hydrodynamic impulses (STAVOPROJEKT Pardubice), or chemical infiltration (VODNI ZDROJE Praha and Holesov, EKOLPraha, ECOLAND Praha, AQUATEST Praha, GEOTEST Brno, etc.).

A: The extracted product can be subsequently recycled.

R: The extracted groundwater can be cleaned in bioreactors (EKOL and ECOLAND Praha) or by other means and discharged to the subsurface.

HYDROGEOLOGICAL METHODS: Venting (B.2)

P: Venting depends on the volatility of some OP, which can be increased with a change of pressure conditions. The OP are then pumped out along with both water and air, and subsequently absorbed by a sorbent material, such as activated charcoal, chemical sorbent, etc.

E: A venting system was first used by STAVEBNI GEOLOGIE Praha. The price varies from $300-17,000 (USD) per cubic meter for chlorinated hydrocarbons.

D: High costs and technical problems in the case of heterogeneous aquifers.

A: The application is most effective in the case of concentrated chlorinated hydrocarbon pollution of an aeration zone.

R: This technology also includes a stripping method in which OP are released from the extracted groundwater by airflow.

BIOTECHNOLOGY: Stimulation of natural biodegradation processes by artificial nutrients and other materials (C.1.a)

P: This method consists mainly of applying artificial nutrients to stimulate biodegradation.

E: This method is used by EKOL Praha, VUAB Praha, ECOLAND Praha, ENVISAN Ceske Budejovice, BIOSAN Praha, VSCHT Praha, and others. Prices are comparatively low and vary from about $1-15 (USD) per cubic meter of soil processed.

D: This method was developed especially for the remediation of soils. Use of this method is patented and is accepted professionally.

A: This method is applied *in-situ* and is considered to be without adverse side effects if correct dosing is used. Basic nutrients are readily available and no special tests are needed for their application.

BIOTECHNOLOGY: Stimulating natural biodegradation processes by natural nutrients (C.1.b)

P: This method involves the application of peat, manure, or similar materials.

E: Peat has been used by RASELINOVE ZAVODY Praha and AREA Praha, manure by STAVEBNI GEOLOGIE Praha and KAP Praha, fermented animal waste by AGROTEM Teplice and AGRONOM Praha, and rinds and other organic remains by ALPHA BIO Praha. Prices vary from $20-100 (USD) per cubic meter of soil.

D: This method can be used *ex-situ* only. A relatively large amount of nutrients is needed. The nutrients are mixed with the polluted soil in proportions of 1:1, though sometimes up to 1:5. The biodegradation process requires a comparatively long time, from seven months to three years. In the case of massive pollution, the method may not be effective. The basic disadvantage of the method is that some nutrients (e.g., peat) are non-renewable. the use of which is not legal in some countries (e.g., Germany).

A: This method can be used even by a non-specialized company because it is not necessary to have any permission or licenses from the authorities for its application. The end product can be used, depending on the final pollutant levels, for agricultural purposes.

BIOTECHNOLOGY: Stimulating natural biodegradation processes by bioventing (C.1.c)

P: Stimulation of natural biodegradation processes is caused by intensifying air exchange in soils.

D: Comparatively high costs.

A: The method is effective and unique in the case of pollution of an aeration zone.

BIOTECHNOLOGY: Application of allochthonous bacterial strains (C.2.)

P: A strain separated from contaminated soils and waters is used.

M: The method was introduced by EKOL Praha, VUAB Roztoky, VUGI BIOSAN Praha, ENVISAN Ceske Budejovice, VSCHT Praha, ALPHA-BIO Praha, AQUA PLUS Praha and is applied by ECOLAND Praha, LINEK and DEKONTA Kladno, DEKOS Hradec Kralove, AGRECO and ECOSYSTEM Praha, NEPTUN Plzen, and others. A cost of less than $20 (USD) per cubic meter is typically encountered.

D: Licenses are needed.

A: High-speed biodegradation lasting up to only three months.

BIOTECHNOLOGY: Application of autochthonous bacterial strains (C.3.)

P: The method applies separated bacterial strains directly to the pollution source.

E: The method is principally used by APV Praha (an agency of UMWELTSCHUTZ Germany), VANKO Liberec, OPV Praha and VSCHT Praha, etc. Prices vary from $2.5-50 (USD) per cubic meter.

D: This method places a very high demand on laboratory equipment and on homogenization arrangements. The method is temporally exacting and expensive. A special license for application of each separate bacterial strain is necessary.

A: There are a minimum number of objections to the method from an environmental point of view because the strains are of natural origin.

COMBUSTION

P: Combustion is one of the "classical" methods. For remediation of the Soviet installations, use of mobile incinerators has been suggested.

E: Construction of special incinerators is planned for this purpose. The price without transportation varies from $50-600 (USD) per cubic meter depending on the type of hydrocarbons. For a common fuel the price is about $35 (USD) per cubic meter of soil.

D: Licenses from the Czech Air Inspection are required. High prices, difficult transport and material sorting are other problems. The method can be used only for remediation of solid or compacted waste with high oil product concentrations.

A: Investment in the process may be low because it is possible to use existing incinerators.

SOLIDIFICATION

P: Waste polluted by OP is stabilized using a material which prevents the OP from being released into the surrounding medium. The method was mainly applied *ex-situ*; the *in-situ* variant was first applied by ECOLAND in cooperation with GEOINDUSTRIA Praha.

E: Recently, BENZINA has specialized in this method since it has its own licensed solidification arrangement. This method is being studied for possible use by other companies, including, VUGI Brno, AGRECO Praha, HYDROGEOLOGIE Chrudim, GEO-ING Jihlava, ECOLAND Praha. The prices vary from $30-300 (USD) per cubic meter of soil.

D: This method requires special arrangements and the use of special remediation methods for different types of wastes. The solidifiers need to be licensed.

A: A large amount of waste can be quickly eliminated.

STORAGE/Disposal

P: Waste is stored/disposed at an environmentally secure landfill.

E: Storage/disposal is a classical method of remediation. The price is about $10 (USD) per cubic meter of soil.

D: Facilities for extremely toxic waste are not available everywhere.

UNSPECIFIED METHODS

R: When proposing to remediate former Soviet installations, some companies have not specified the technologies they intend to apply.

7. Problems with Remediation Activities at Former Soviet Installations

Most problems related to remediation activities for these sites involve remediation criteria, bidding and award procedures, financing, proposed remediation technologies, and monitoring and supervision.

7.1. Remediation Criteria

Many remediation activities have been performed without any prior risk analysis because most Czech companies have not had experience using it. As late as 1990, official Czech clean-up standards were not yet available. In 1991, a limit value for OP of 100 mg kg^{-1} of dry residue was set. Based on the so called Holland Agreement, the limit value was increased to 1,000 mg kg^{-1} in 1992. The increased limit values resulted in a significant decrease in the cost of remediation projects. Local authorities have, however, been interested in maximizing remediation works regardless of the real environmental hazard (in order to get maximum financial support from the state budget). The fact that most companies and local authorities have not yet applied the principles of risk analysis in remediation projects remains a major unsolved problem and a significant need related to the remediation of the former Soviet installations throughout the region.

7.2. Bidding and Award Procedures

The procedure for bidding and evaluation of proposals was developed by a special governmental office in the Czech Republic. Objectivity of the procedure was guaranteed by the "Commercial Bureau." Projects were reviewed and costs estimated by a professional commission. The names of the commission members were not made public. The members were representatives of the special government office, the former federal Czech and Slovak

Ministries of the Environment, as well as outside experts. Both local and foreign companies could take part in the competition.

A majority (94%) of projects were originally granted to four hydrogeological companies (SG Praha - 48%, GEOTEST Brno - 26%, NEPTUN Chrudim - 11% and VZ Praha - 9%). The rest (6%) were divided among five other companies (VPU Praha, UNIGEO Ostrava, IG Ostrava, EKOL Praha and VZ Holesov). Only nine companies were selected for the first stage of remediation at former Soviet installations in the Czech Republic. Unfortunately, winning projects were not reviewed by other professionals. To improve this procedure, it recommended that future competitions be organized into two stages: an initial competition involving selection of the appropriate remediation technology (i.e., strategy, remediation techniques, etc.), followed by the submission of plans regarding completion of the remediation work (i.e., costs, time schedule, guaranties, etc.).

7.3. Financing

Contrary to the original belief that the environmental remediation of these sites would be paid for by the Soviet government, remediation costs are instead paid for by the Czech state budget. Contracts terms have not been based on completed remediation, but instead on year-to-year funding availability. This may be an advantage for the companies, and a disadvantage for the State, because the projects usually do not guarantee a final result within a particular time frame. This allows some companies to apply a principle of "maximum financial support now, and then we shall see".

7.4. Proposed Remediation Technologies

Until 1991, companies in the Czech Republic primarily had experience with remediation pumping (SG Praha, GEOTEST Brno, NEPTUN Chrudim, and others) and. therefore, gave priority to this technology. More recently, thanks to progress with biotechnologies (introduced on the Czech market by EKOL Praha), and venting (AQUATEST Praha), it is now possible to employ alternative technologies. Given the long delays (sometimes lasting longer than one year) between project submission and the beginning of remediation, many problems have been encountered, for example:

- remediation limits were changed (usually increased, e.g., for non-polar hydrocarbon extraction, levels were raised from 100 to 1000 mg kg^{-1} of solids);

- original requests for the demolition and reconstruction of some installations were never completed;

- in the case of near surface pollution, petroleum products may have evaporated, while in the case of groundwater pollution, a pollution plume may have spread;

- many of the former Soviet military installations were sold or leased to new "users";

- some projects were developed with the assumption that damages would be paid for by the Soviets; and

- decontamination limits at a majority of the sites were fixed regardless of the future use of the sites (parking lots for cars and tanks, etc.). In the case of groundwater pollution, the limits were determined according to Czech standards regardless of background pollution in the area.

Now that Czech companies have become familiar with remediation techniques and have gained significant experience, they are able to compete more effectively with foreign companies. A major concern, however, is that some companies may apply technologies regardless of their actual effectiveness.

7.5. Monitoring and Supervision

Remediation monitoring and supervision of the remediation activities is conducted through use of control samplings. The costs of the supervision are high, often comparable with the costs of the remediation itself. Specific procedures and standards for the supervision of remediation results, as well as for monitoring, have yet to be set. These procedures are, of course, not as important for the companies as they are for the local municipal authorities.

8. Conclusions and Recommendations

Experience in solving the environmental problems at former Soviet military installations in the Czech Republic is of great importance. Addressing these problems has provided Czech firms with invaluable experience in the application of different remediation technologies and approaches. The experience can be used not only for the remediation of the former Soviet installations, but also in the remediation of polluted civil or industrial sites.

Based on nearly five years of experience, the establishment of a common financial fund is recommended. In this proposed fund, EKOFOND, both foreign aid and state budget funds (including funds from the Ministry of Environment, Ministry of National Defense, and others) would be combined for site remediation activities.

As a fundamental condition for improved remediation effectiveness, remediation results paid for using public funds should be readily accessible by the public and openly published.

References

ECOLAND (1994): Archived material of ECOLAND Prague.

Svoma, Jan (1991): Hydrogeological survey of rock and groundwater pollution caused by the Soviet army stay in Czechoslovakia. *Geologicky pruzkum*, 7, 1991, 198-200 (in Czech).

ENVIRONMENTAL PROBLEMS AT MILITARY INSTALLATIONS IN

GERMANY

Dietrich Burkhardt
Industrieanlagen-Betriebsgesellschaft mbH
(IABG)
Einsteinstr. 20
D-85521 Ottobrunn
Germany

Introduction

Up until 1990, approximately 1 million hectares of land have been used for military purposes in Germany (see Figure 1). About 500,000 hectares will be handed over to civil agencies by the middle of the nineties. At the beginning of 1991, the Federal Ministry of the Environment, Nature Conservation and Nuclear Safety, through the Federal Environmental Agency and in agreement with the Federal Ministry of Finance, commissioned Industrieanlagen-Betriebsgesellschaft mbH (IABG) to conduct an inventory of suspected contaminated areas on the West Group of the Soviet Troops (WGT) properties in Germany. This project has been carried out under the general management of IABG together with about 30 partner companies from the new Federal States.

Approximately 1,200 sites, with a total area of about 270,000 hectares, have been investigated. These properties vary significantly in terms of usage and size. The various types of usage include troop training areas, airfields, fuel tank farms and barracks. These are also the properties with the greatest area and, presumably, with the greatest contamination.

Figure 2 shows the type and distribution of military properties in the state.

A large portion of the properties are situated in the Federal States of Brandenburg and Saxony-Anhalt. In Brandenburg alone, there are 339 properties comprising about 40% of the total area. The properties are to be vacated by the end of August of 1994 based on the joint statement issued by the German Chancellor and the Russian President on December 16, 1992.

NATO ASI Series, Partnership Sub-Series, 2. Environment – Vol. 1
Clean-up of Former Soviet Military Installations
Edited by R. C. Herndon et al.
© Springer-Verlag Berlin Heidelberg 1995

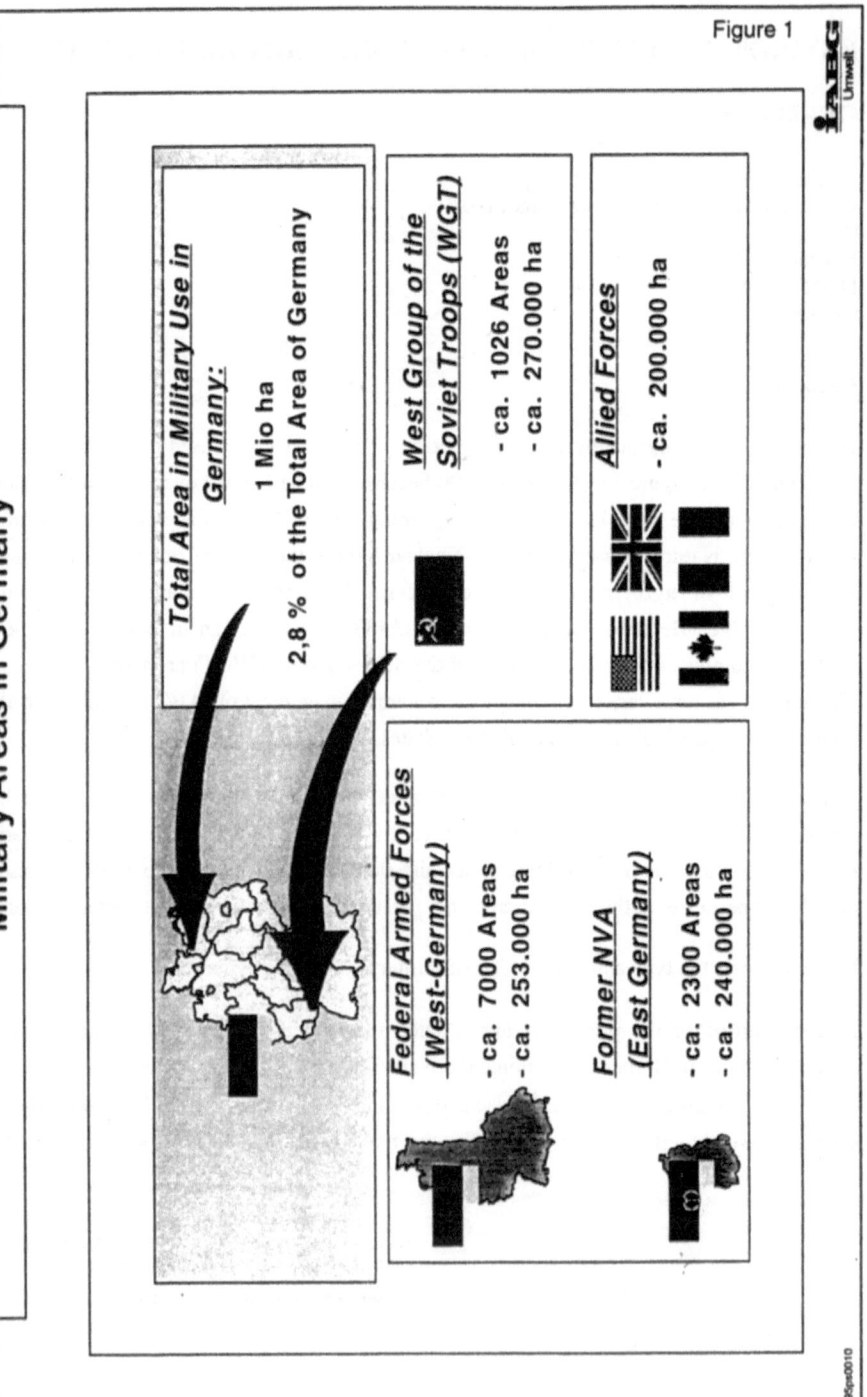

Figure 1

Type and Distribution of the WGT Properties

Figure 2

	Berlin	Branden-burg	Mecklenburg Pomerania	Saxony	Saxony-Anhalt	Thuringia	Total
Administrative buildings	0	14	10	8	11	5	48
Barracks, quarters	7	137	34	72	109	58	417
Training facilities	1	4	1	0	1	2	9
Telecommunication facilities	0	31	11	13	23	15	93
Airfields	0	29	9	11	26	5	80
Depots, stores, bunker	0	35	11	16	31	24	117
Troop trainings areas	0	43	21	24	65	20	173
Repair workshops	1	16	3	5	1	2	28
Other facilities	2	30	4	10	10	5	61
Total number	11	339	104	159	277	136	1026

IABG
Umwelt

95ps0001

After withdrawal of the troops, the properties are handed over to the responsible Federal Property Agency in the new Federal States, and become part of the general Federal assets. The governments of Saxony, Thuringia and Brandenburg have accepted the offer of the Federal Government to take over almost all sites which are not to be used by Federal institutions (e.g., the "Bundeswehr"). After taking over these sites, the costs of remediation are to be paid by the respective State.

According to the troop withdrawal contract, the site-related processing of the properties can take place after notice to vacate has been received by the relevant Federal Property Agency. This notification should be given two months prior to the withdrawal in question. In practice, however, the time allowed is typically shorter than two months.

Objectives of the Project

The objectives of the project are to:
1. screen, record, describe and document suspected contaminated areas on the properties;
2. determine and report dangerous situations and submit proposals for implementing emergency measures to avert immediate danger to the responsible authorities;
3. conduct a comparative initial evaluation of all suspected contaminated sites; and
4. carry out risk assessments for selected properties in order to establish relevant criteria for further treatment of the sites.

Structure of the Project

The project is divided into a number of components (see Figure 3). Typically an inventory and compilation of basic data are made; aerial photographs have been taken and analyzed for all sites. Specific on-site determination of the suspected contaminated areas, as well as initial evaluation is conducted and emergency measures are established.

The project is monitored by a project advisory committee. The following institutions are represented:

> The Federal Ministry of Environment, Nature, Conservation and Nuclear Safety;
> The Federal Ministry of Finance;
> The Federal Ministry of Defense;
> The Federal Ministry for Regional Planning, Building and Urban Development;
> The Federal Environmental Agency; and
> State Ministries of Environmental Protection in the new Federal States.

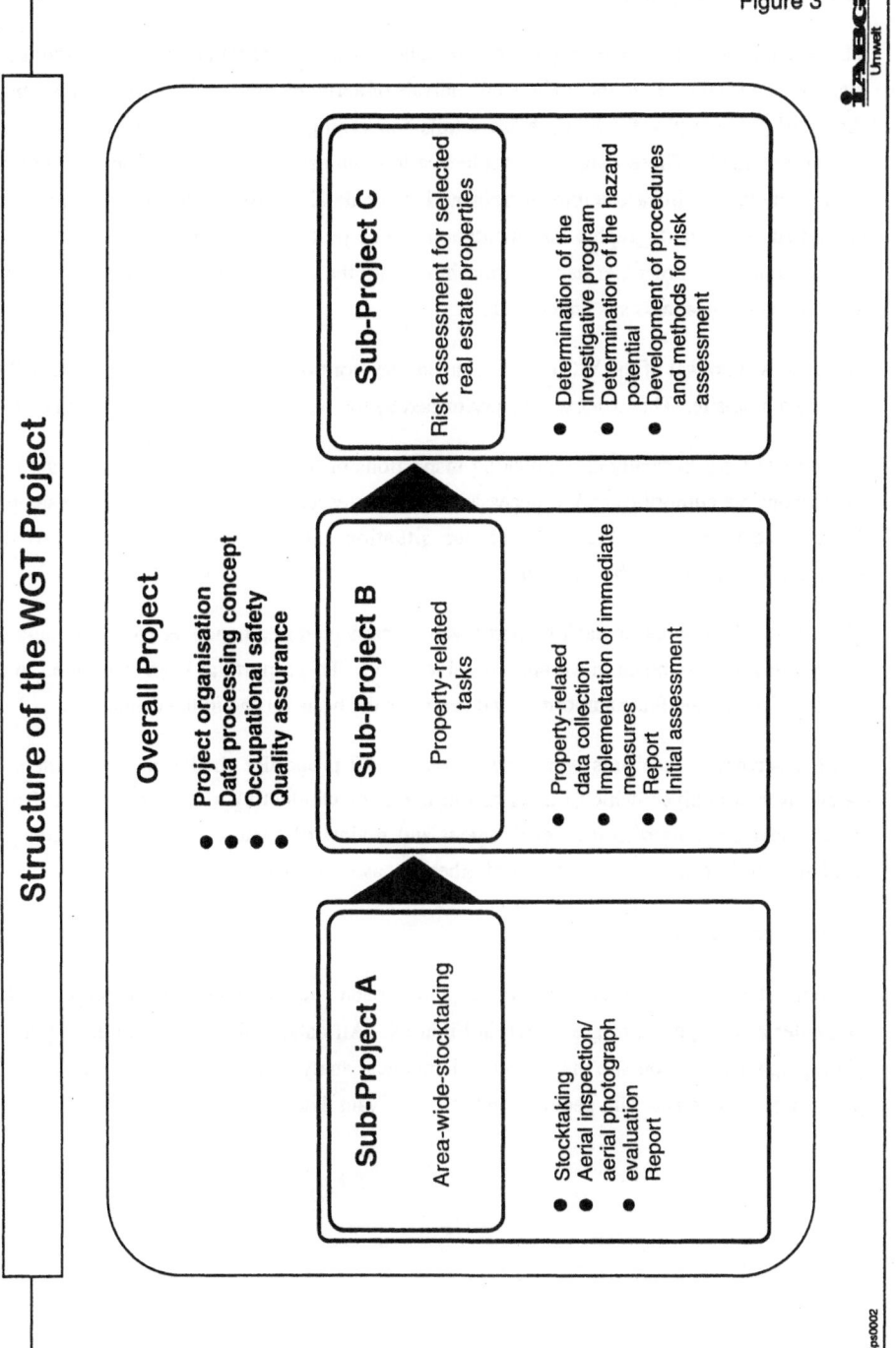

Structure of the WGT Project

Figure 3

Overall Project

- Project organisation
- Data processing concept
- Occupational safety
- Quality assurance

Sub-Project A

Area-wide-stocktaking

- Stocktaking
- Aerial inspection/ aerial photograph evaluation
- Report

Sub-Project B

Property-related tasks

- Property-related data collection
- Implementation of immediate measures
- Report
- Initial assessment

Sub-Project C

Risk assessment for selected real estate properties

- Determination of the investigative program
- Determination of the hazard potential
- Development of procedures and methods for risk assessment

iABG Umwelt

95ps0002

Procedure and Acquisition of Data

The relevant basic data for each property are collected at the earliest possible time. Particular significance is attached to the taking and analysis of aerial photographs. Through analysis of the aerial photographs, the first precise details concerning the suspected contaminated areas can be obtained. These data are recorded cartographically, along with other topographic details, including infrastructure, terrain and plant details. By using modern evaluation equipment, exact cartographic documentation can be produced. In the event that properties were used by the former German Wehrmacht up until the end of World War II, multitemporal aerial photograph analysis is undertaken.

Once the withdrawal of the WGT is announced, the work on the properties concerned will be undertaken on-site. This work will be performed by the partner companies mentioned earlier.

The on-site tasks basically entail making inspections of the properties, making contact with the responsible authorities and, if necessary, for the preservation of evidence, taking samples for chemical analysis. Acute hazardous situations are reported immediately so that emergency measures can be initiated.

The results of the work on each property were documented in a handover record through the end of 1992 and are noted in a comprehensive report. This report has been distributed to the relevant Federal and State authorities and serves as the basis for the initial evaluation.

Further screening measures were carried out on 20 properties, including comprehensive sampling and analysis which serves as the basis for conducting risk assessments. These results have been used in the development and review of basic methods, procedures and guidelines for the economic treatment of other risk assessments or evaluations.

Procedures and Methods

A comprehensive data processing concept has been implemented for the project. The computer-aided operations are shown in Figure 4. All data, which were acquired by aerial photograph analysis, on-site inspection and chemical analysis, are recorded by the ALADIN program and combined with geologic and other relevant data.

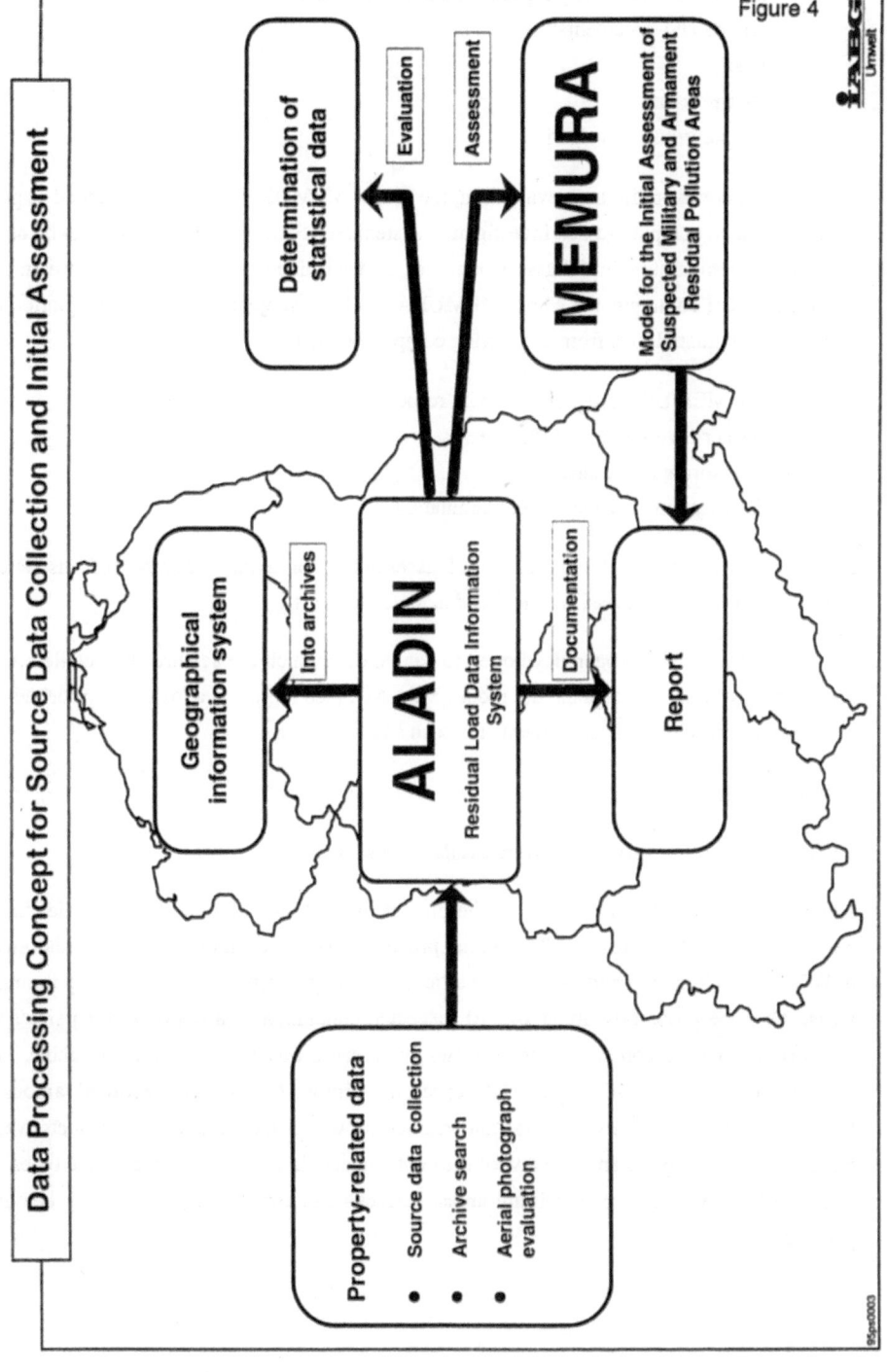

Data Processing Concept for Source Data Collection and Initial Assessment

Figure 4

ALADIN provides the necessary input data and procedures for:
- reports and documents;
- maps;
- statistics; and
- the preliminary evaluation.

In order to obtain a preliminary evaluation, the model MEMURA has been developed as part of the integrated computer-aided investigation system (see Figure 5). MEMURA can be used to achieve a comparative initial assessment, i.e., to determine priorities for further treatment of the suspected contaminated sites. MEMURA has been subject to a special adaptation to military contaminated sites from an existing computer model.

The results of MEMURA are priorities with respect to:
- further investigation of the sites;
- monitoring of the sites; or
- falsification of suspected contamination.

Figure 6 shows the results of an initial assessment for a large number of suspected contaminated sites on the basis of the 5 endangered elements.

The model ALADIN has been developed to handle data which are acquired by detailed site investigations. Using these data, the model MAGMA (see Figure 7) allows new priorities to be set using individual risk assessments for each site.

Preliminary Results

The following preliminary results were available in May of 1994.

Altogether, about 1,000 properties on the project list and an additional 90 properties have been assessed totally or in part. The partial processing is the result of the partial withdrawal of WGT over a longer period of time, in particular for larger properties, such as troop training areas. Of these properties, about 24,110 suspected contaminated areas of varying size and intensity have been recorded. Figure 8 shows that mineral waste such as building rubble, ash or contaminated excavation material represents almost half of the recorded amount. Similarly, larger quantities of metal and residential waste and mineral oil products were recorded. Figure 9 gives an overview of the contaminants in tons with respect to the types of areas. It is shown that troop training areas, barracks and airfields are particularly heavily polluted.

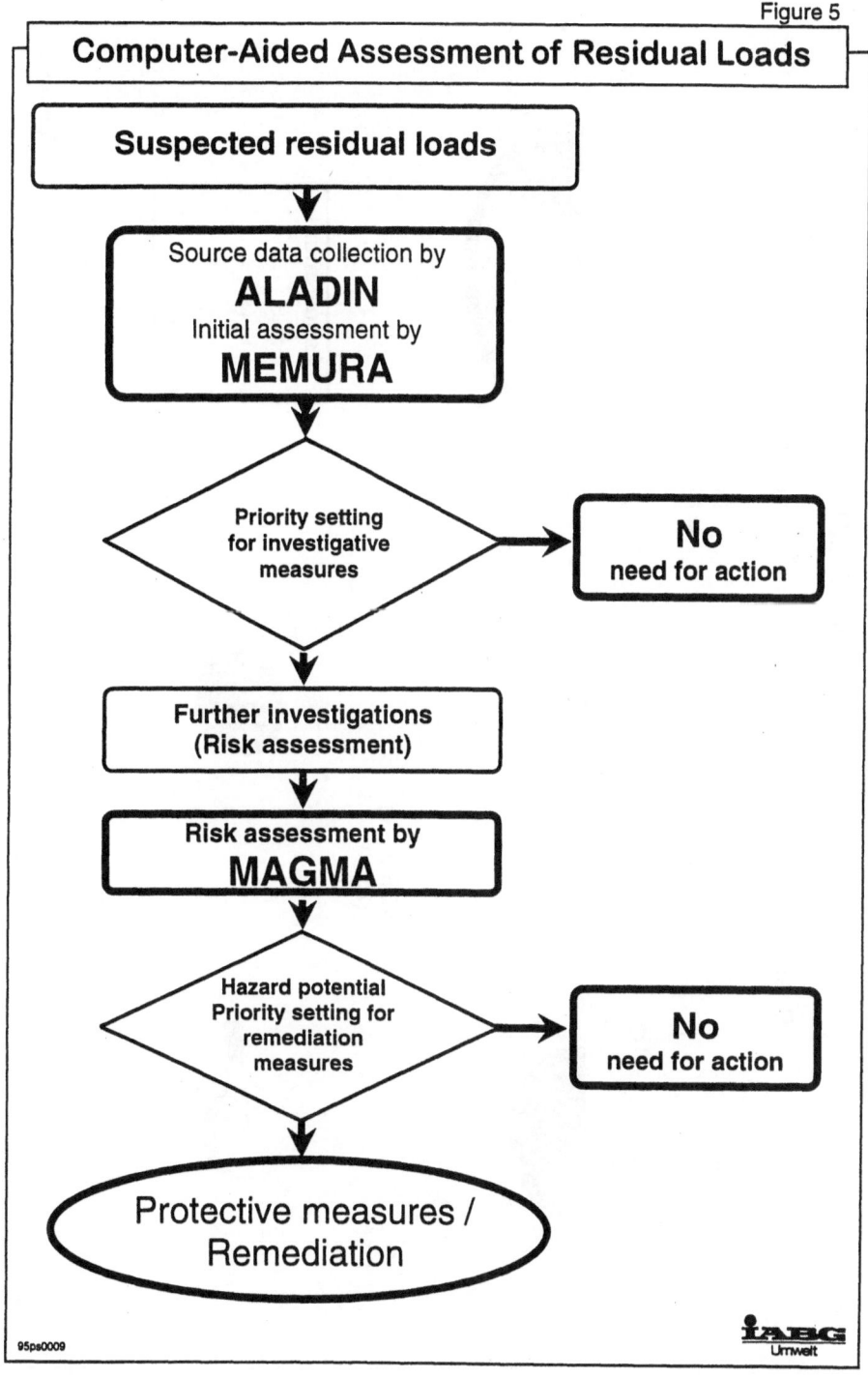

Figure 5

Computer-Aided Assessment of Residual Loads

Suspected residual loads

Source data collection by
ALADIN
Initial assessment by
MEMURA

Priority setting
for investigative
measures

No
need for action

Further investigations
(Risk assessment)

Risk assessment by
MAGMA

Hazard potential
Priority setting for
remediation
measures

No
need for action

Protective measures /
Remediation

95ps0009

iABG
Umwelt

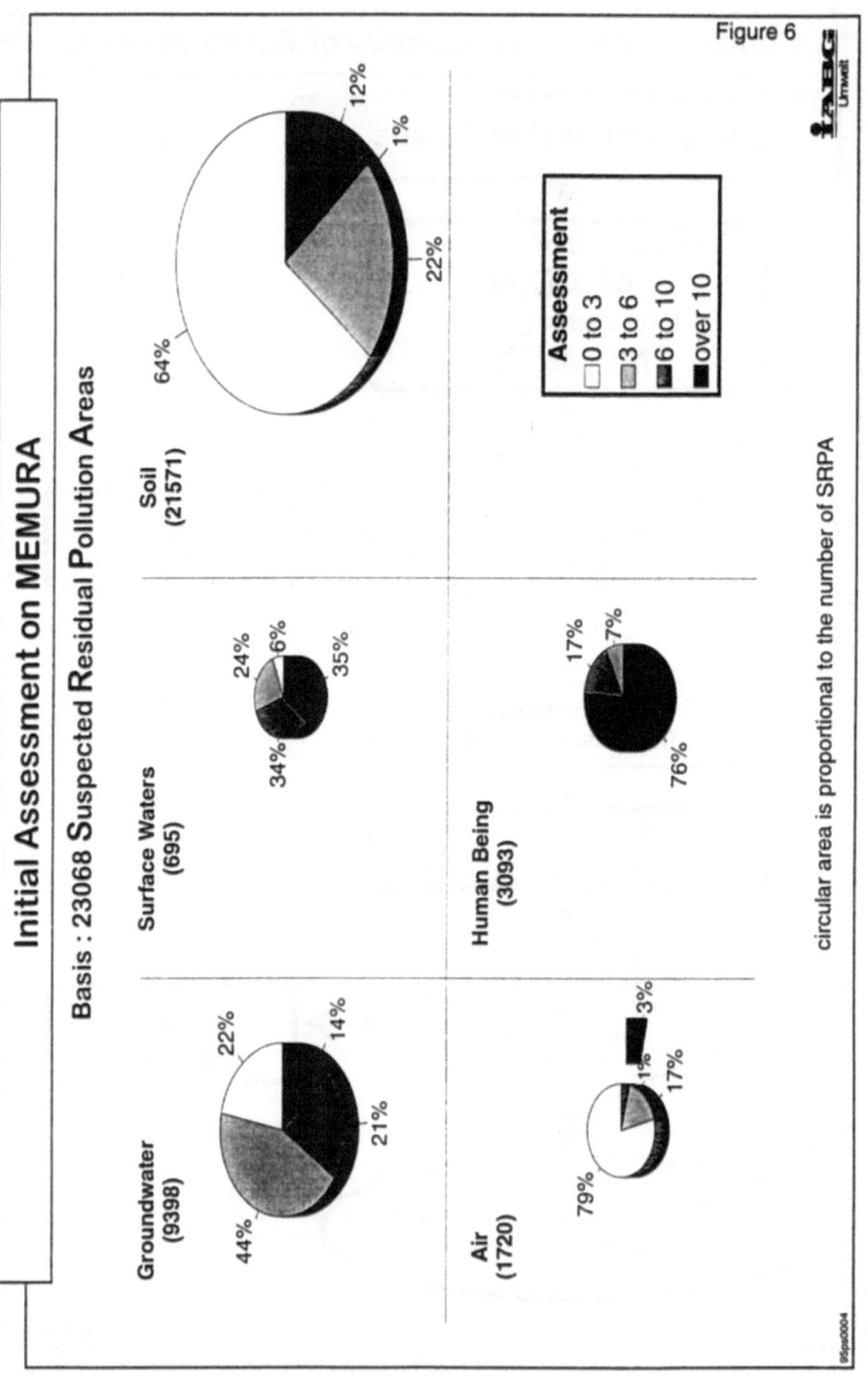

Initial Assessment on MEMURA

Basis : 23068 Suspected Residual Pollution Areas

Soil (21571)
- 64%
- 12%
- 1%
- 22%

Surface Waters (695)
- 24%
- 6%
- 35%
- 34%

Human Being (3093)
- 17%
- 7%
- 76%

Groundwater (9398)
- 22%
- 14%
- 21%
- 44%

Air (1720)
- 79%
- 17%
- 1%
- 3%

Assessment
- ☐ 0 to 3
- 3 to 6
- 6 to 10
- ■ over 10

circular area is proportional to the number of SRPA

Figure 6

Umwelt

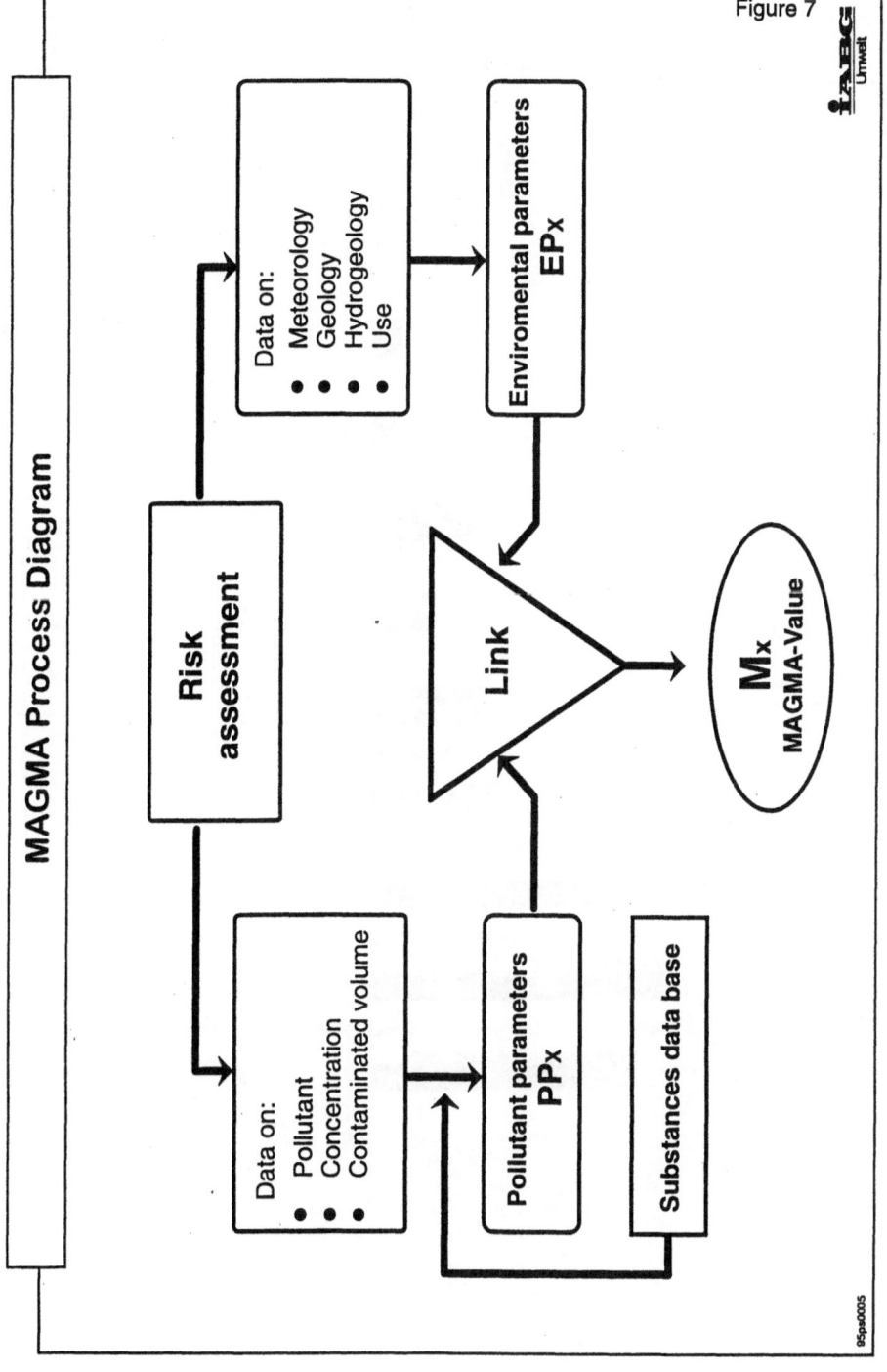

Figure 7

MAGMA Process Diagram

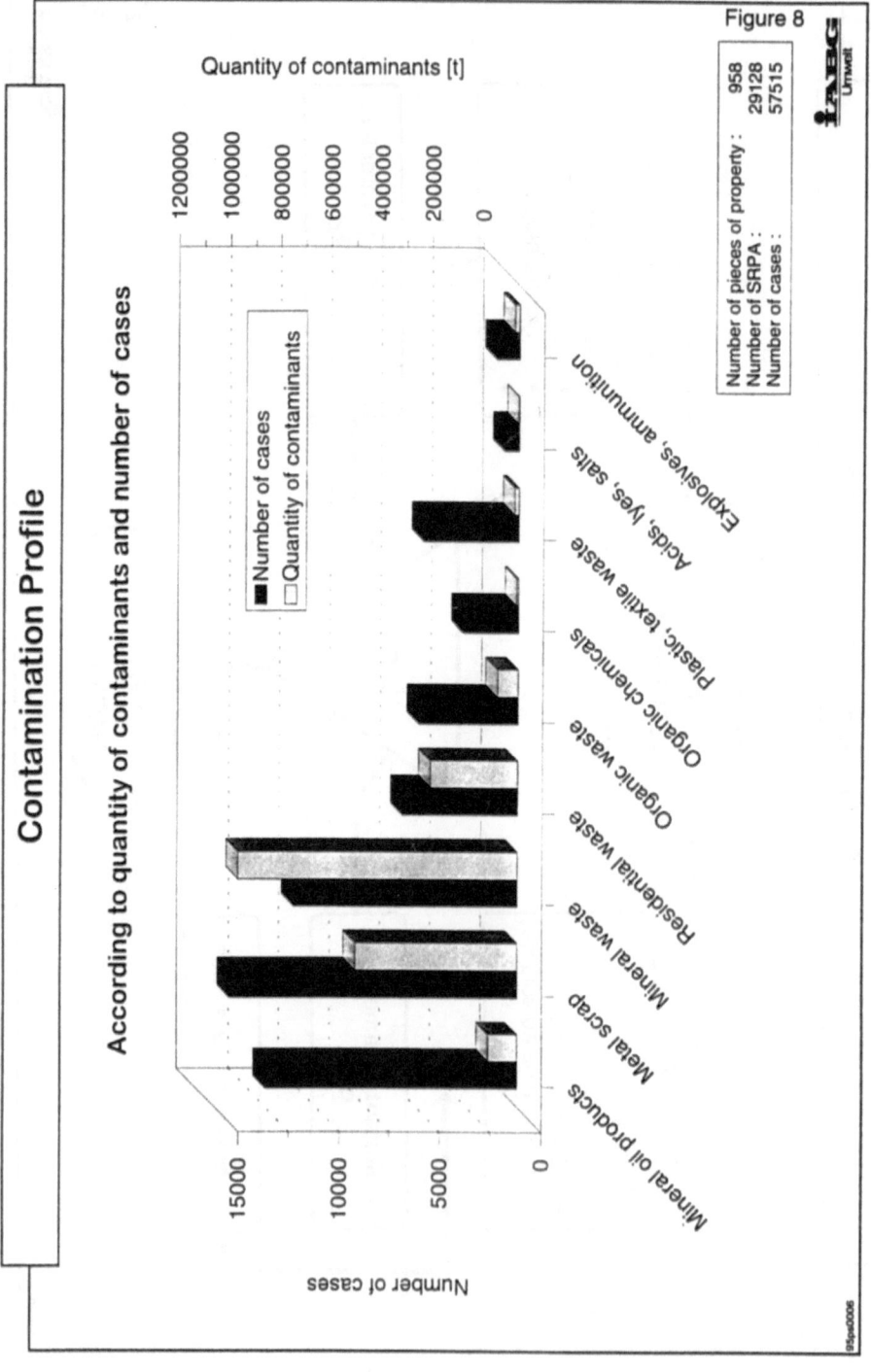

Figure 8

Contamination Profile

According to quantity of contaminants and number of cases

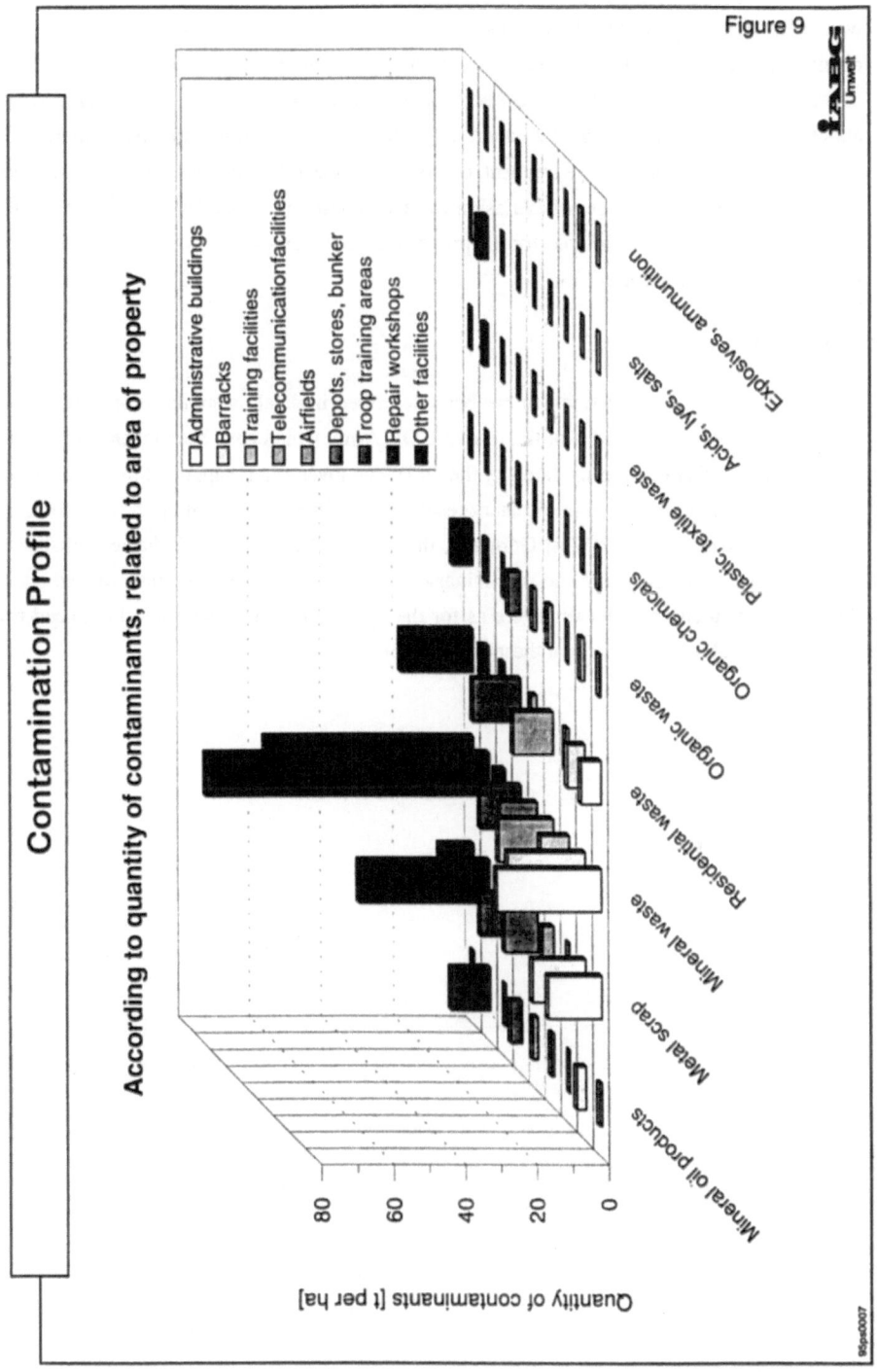

Figure 9

Contamination Profile

According to quantity of contaminants, related to area of property

The necessary course of action will depend on the type, intensity and amount of pollution, as well as geological and hydrological factors and other environmentally relevant boundary conditions according to time, scale and type of the measures determined. In cases where contamination of the soil exists, further screening measures are required. For 445 properties, about 2,500 emergency measures to avert acute danger have been initiated (see Figure 10). Acute risks to water and soil are frequent, comprising some 1,250 reports. Further acute risks exist where there is the danger of explosion from ammunition - about 470 cases. The results of risk assessments are available for 12 properties. Risk assessments for an additional 8 properties have been initiated.

Summary and Outlook

Apart from the WGT project, which has been described in detail, a similar project has been established for those properties of the former GDR Army which have been taken over by the Bundeswehr. This project is operated by the OFD Hannover and is supported by a consulting company. At the moment, no preliminary results are available. Concerning the properties of the Allied Forces in former West Germany, the State of Rheinland Pfalz has established a project to make an inventory and preliminary assessment of suspected contaminated sites. Although the situation is not as severe as for the WGT properties, there will be a need for extensive remediation measures at these sites as well.

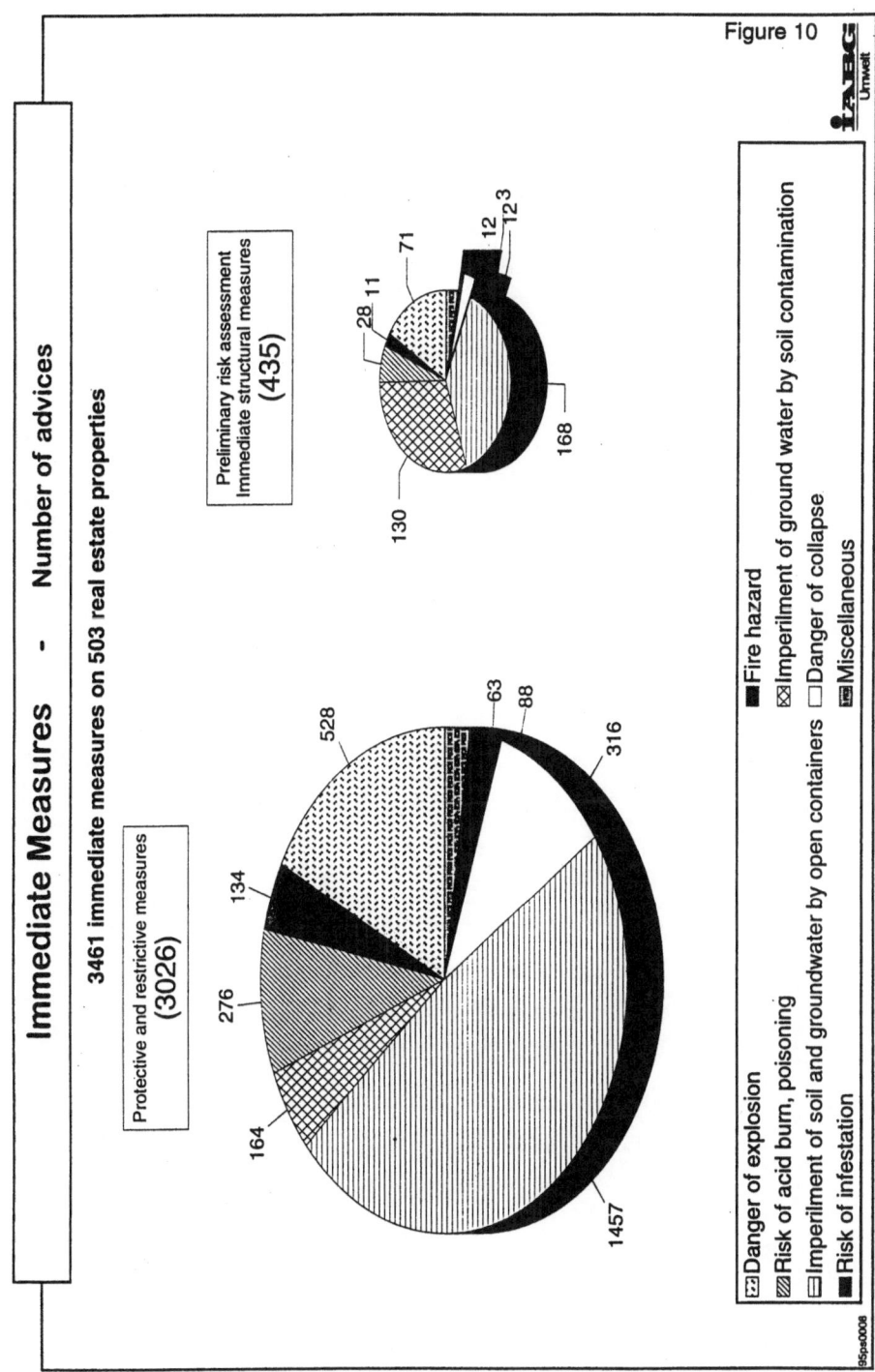

Figure 10

Immediate Measures - Number of advices

3461 immediate measures on 503 real estate properties

Protective and restrictive measures (3026)

Preliminary risk assessment
Immediate structural measures (435)

Danger of explosion
Risk of acid burn, poisoning
Imperilment of soil and groundwater by open containers
Risk of infestation
Fire hazard
Imperilment of ground water by soil contamination
Danger of collapse
Miscellaneous

CLEANUP STRATEGIES FOR U.S. ARMY, EUROPE, MILITARY BASES

Armand C. LePage
Headquarters, U.S. Army, Europe, and Seventh Army
Office of the Deputy Chief of Staff, Engineer
Unit 29351
APO AE 09014
Germany

Abstract

Contaminated sites located on European military installations of the U.S. Army, Europe, are not treated the same as contaminated sites on U.S. Army bases in the United States. On U.S. Army, Europe bases, cleanup is generally triggered by generic numerical criteria, such as the "Dutch list," and cleanup standards are generally set by host nation environmental officials. Although the U.S. Army, Europe does not now establish any priority in funding cleanup of these sites, policy is changing to favor the use of a screening model to allow priority funding of contaminated site cleanup under a "worst first" policy.

Background -- Contaminated Sites on U.S. Army Bases

The following is information on U.S. Army, Europe cleanup of contaminated sites on its military installations. In order to put the strategies of the U.S. Army, Europe into perspective, let me first provide a brief overview of the U.S. Army contaminated site program in the United States.

The U.S. Army operates or has operated installations that manufactured military supplies, conducted various industrial operations, conducted training, housed troops, maintained combat equipment, and did all the other things a modern army must do to stay operationally ready. The most complex and potentially most harmful contaminated sites occur on military installations involved in manufacturing and industrial operations. Those operations were conducted almost exclusively in the U.S., and not in European countries or other countries

NATO ASI Series, Partnership Sub-Series, 2. Environment – Vol. 1
Clean-up of Former Soviet Military Installations
Edited by R. C. Herndon et al.
© Springer-Verlag Berlin Heidelberg 1995

with U.S. bases. Consequently, by far the greatest number of such sites are in the United States.

In order to respond to those contaminated military sites in the U.S., the U.S. Department of Defense (DoD) operates an Installation Restoration Program, or IRP. It is the DoD equivalent of the U.S. Environmental Protection Agency's (EPA) Superfund program. The U.S. Army has run the IRP on its sites since 1975, and has about 1,400 properties nationwide under the IRP program. The steps in the IRP process are familiar by now: Preliminary Assessment/Site Inspection (PA/SI); Remedial Investigation/Feasibility Study (RI/FS); and Remedial Action.

The actions involving cleanup at military sites on EPA's National Priority List (NPL) follow that approach to cleanup, using applicable, relevant and appropriate federal and state requirements (called ARARs in the U.S.), coupled with a site-specific risk assessment. That process does not always lead to cleanup.

An NPL site at the Louisiana Army Ammunition Plant in Shreveport, Louisiana, is illustrative (U.S. Army, 1994). The shallow groundwater at the site is contaminated with chemicals, including explosives. The U.S. Army has proposed a solution involving monitoring and land restrictions (institutional controls) instead of a pump-and-treat solution. The natural degradation of the contaminants is estimated to take 1,600 years; the most rapid pump-and-treat solution would take 400 years. The cost of the selected alternative was $1.3 million (USD), versus $6.1 million (USD) for the least expensive pump-and-treat system. The rationale given was that reasonable exposure scenarios were unlikely to occur (no shallow drinking water wells would be drilled, and thus there would be no receptor population) or would not result in damage (exposure from contaminated groundwater discharge into nearby streams would not create unacceptable risk to recreational uses). The alternative recommended by the U.S. Army has not yet been approved by either EPA or Louisiana environmental officials.

U.S. Army Contaminated Sites in Europe

The following is information concerning the contaminated sites of the U.S. Army bases in Europe. In early 1990, the U.S. Army maintained over 800 installations in Europe. That number has drawn down to less than 400 over the past 4 years. On those installations, there are approximately 240 contaminated sites.

There are two significant differences between contaminated sites on Army bases located in the U.S. and Army bases located overseas. First, very few if any overseas installations conducted manufacturing or industrial operations; therefore, these tend to be limited to less complex contamination problems. For example, almost all of the contamination in the U.S. Army, Europe bases originate from spills or leakage of fuel oil or. solvents, such as chlorinated hydrocarbon compounds. Second, there is no special program, nor special funding, dedicated to cleanup of U.S. installations overseas as there is for installations in the U.S. Thus, while the U.S. Army, Europe has much less severe cleanup problems, neither the EPA Superfund program nor the DoD Installation Restoration Program control or fund the cleanup process at these sites. Instead, the cleanup processes in Europe are governed by a combination of Host Nation law and international agreements, and processes devised by U.S. Army, Europe. Specifically, the requirements of Host Nation law and international agreements, such as those among the NATO nations, largely control the cleanup levels; the actual process of planning, funding, and management of a program to cleanup U.S. Army contaminated sites in Europe is the responsibility of the U.S. Army, Europe.

Cleanup at U.S. Army, Europe Installations

The basic question in the cleanup decision process on U.S. Army, Europe installations is "how clean is clean." The issue of "how clean is clean" is often framed as a choice between using generic numerical criteria, such as the "Dutch list," or a site-specific risk assessment approach. In practice, the generic criteria approach has often been coupled with a general cleanup level to achieve soil "multifunctionality." The site-specific risk assessment, more common in the U.S, has usually been coupled to a cleanup level matched to the intended future use of a specific site. While those approaches are often used in tandem today -- generic criteria in early stages of assessment, and site-specific methods at the point of decision making (Visser, p. 2) -- the U.S. Army, Europe (especially in Germany) has in the past generally considered the question of cleanup in terms of generic numerical criteria, rather than in terms of site-specific risk-assessment. The reasons for this are generally legal and practical, including the international agreements governing the U.S. Army in Germany, the role of the U.S. Army as tenant, and the U.S. Army policy of environmental compliance. Each of these reasons is briefly described in the following paragraphs.

International Agreements: The NATO States that have forces stationed in Germany are signatories to the NATO Status of Forces Agreement, Supplementary Agreement (SOFA SA). For purposes of environmental compliance, the SOFA SA has been interpreted by the U.S. Army, Europe to mean that it will follow substantive German law. The revised SOFA SA, which is likely to come into effect in 1994 or 1995, strengthens that interpretation. Thus, in the first case, the SOFA SA acknowledges German legal authority in issues of environmental compliance. Site-specific risk assessment must be considered such an area, particularly since German citizens are likely to be affected.

U.S. Army, Europe as Tenant: The U.S. Forces are generally tenants on the land owned by the German Federal Ministry of Finance. Thus, decisions bearing on the ultimate use of the land are more appropriately made by the German FMOF.

Environmental Compliance: The U.S. Army provides funds for projects to come into compliance with Host Nation law. Under DoD procedures, non-compliance with Host Nation law is generally accorded the same priority funding as violations of U.S. law. DoD policy is to fund the correction of all non-compliance problems, regardless of the type or severity of the non-compliance. Accordingly, since funding is universal, there has been no reason to establish a priority list of projects. Non-compliance for contaminated sites has generally been based on numerical generic criteria (i.e., concentrations developed at the German State level) or sometimes on the so-called "Dutch list," the A-, B- and C-levels of the Netherlands Soil Cleanup Guidelines. Violation of these State level or the "Dutch list" C-level is generally the signal for non-compliance, thus signaling the need to fund a cleanup project.

To summarize, the international agreements which govern U.S. Forces in a Host Nation and the tenant relationship of the U.S. Army on its European installations requires the U.S Army to defer to the Host Nation in the use of site-specific risk assessment to determine cleanup levels, or priorities. On the other hand, the DoD has a policy of funding projects to correct non-compliance, which in the case of contaminated sites, is generally signaled by C-level or equivalent levels of contamination.

Changes in U.S. Army, Europe, Approach

Funding pressures, including Congressional pressure, have caused the U.S. Army, Europe to reconsider the policy of viewing all non-compliance issues as equally important. In particular, the U.S. Army, Europe is reviewing how to place its cleanup projects in some sort of funding priority.

The U.S. Army, Europe has developed a simple model for establishing funding priorities among its cleanup projects. The model is based on the older Hazard Ranking System (HRS) used in the U.S. EPA's Superfund program and the Defense Priority Model (DPM), used by the DoD for its Installation Restoration Program. It is not intended to function other than as a "screening" device, useful in directing funding to the most dangerous problems. As a screening model, it is greatly simplified in comparison to the HRS and DPM models, principally in order to reduce the data requirements needed to run the model. In particular, the model requires relatively less site characterization data than the U.S.-based models. It also focuses only on the groundwater pathways of human exposure.

The model, called the Prioritization Model, is intended to provide a rational approach to assigning relative ranking among U.S. Army contaminated sites. It is not intended to replace the site-specific risk-assessment role of the Host Nation authorities, who must bear principal responsibility for determining site-specific cleanup standards for soil and groundwater resources in their countries.

The Prioritization Model is incorporated in a larger model, called the Database of USAREUR Contaminated Sites, or DUCS. DUCS was developed by U.S. Army, Europe to augment and extend the information available for contaminated sites reported in the standard U.S. Army-wide environmental funding database, which is used to establish the need and request the funding for environmental projects. DUCS extracts all funding information for contaminated sites from the environmental funding database, and allows for adding technical and management data elements and information necessary to provide appropriate headquarters oversight of the cleanup requirements on U.S. Army installations throughout Europe. The DUCS model compares contaminant levels with site survey results, flagging values that exceed the "Dutch list" C-values. The Prioritization Model is an algorithm run from the overall DUCS model.

The Prioritization Model is divided into three components:
- Potential to Release (PR): Identifying whether the waste has released and penetrated the groundwater or whether the site conditions are such that the waste can easily leach into the groundwater. PR considers measures of net precipitation, depth to aquifer, permeability, and degree of containment.
- Waste Characteristics (WC): Identifying how toxic and mobile constituents are in the waste and the quantity of waste that may be potentially released into the groundwater. The WC portion of the Model groups contaminants into 14 classes, based on toxicity/mobility considerations.
- Potential Affected Population (PAP): Identifying the nearest groundwater well and the size of the PAP served by the well or group of wells.

Sub-scores are derived for each of the above components of the Model. The algorithm then multiplies the subscores, and adjusts the overall score so that it falls between 0 and 100.

The U.S. Army, Europe believes that priority for action in the cleanup of contaminated sites must be risk-based. Since resources are always limited, funding must be directed towards the most threatening problems, i.e., "worst first." Once the priority for funding has been established through an initial screening model (such as the Prioritization Model described above), the more detailed issue of site-specific risk assessments, of determining "how clean is clean," still continue to be the province of the Host Nation in which U.S. military bases are located.

Conclusion

The use of a risk-based screening model, such as the Prioritization Model of the U.S. Army, Europe, provides a readily usable indicator of those projects which should receive priority in funding. Such a screening model cannot answer the question of "how clean is clean", which must be determined through a site-specific risk assessment. Such a model, however, can be the basis for determining how best to allocate funding to minimize risk from contaminated sites.

References

Visser, Wilma JF (1993), Contaminated Land Policies in Some Industrialized Countries. Technical Soil Protection Committee, The Hague.

U.S. Army (1994), Final Proposed Remedial Action Plan, Louisiana Army Ammunition Plant, Shreveport, Louisiana. Environmental Science and Engineering Co., St. Louis, MO.

REMEDIATION FROM 1991 TO 1994 OF ENVIRONMENTAL

DAMAGES CAUSED BY THE SOVIET TROOPS

István Endrédy
General Director
Institute for Environmental Management
Alkotmány u. 29
H-1054 Budapest
Hungary

Abstract

On October 24, 1991, the Hungarian Government ordered an immediate remediation of the pollution at 20 former military bases and prescribed the urgent protection of sensitive areas (e.g., populated areas, drinking water sources). For these tasks the government allocated 930 million Forints (Fts). By the end of 1993, eight of the bases had been fully cleaned up with respect to the known pollutants. Twelve others required more work. During this time, 11.3 million liters of kerosene, diesel oil and petrol were extracted and rendered harmless. Approximately 586,000 cubic meters of polluted groundwater were removed and treated. Finally, 195,000 cubic meters of soil were decontaminated and backfilled under well-controlled conditions.

1. Introduction

In accordance with the intergovernmental agreement of March 10, 1990, environmental damage was assessed at 171 garrisons, totaling an estimated 60 billion Fts. The total area of the pollution was about 48,000 hectares across 340 settlements within the country. This assessment was carried out from December 20, 1990, to June 10, 1991. The damage assessment methodology was reconciled with Soviet experts between September 1991 and February 1992. The cost of the assessment was 130 million Fts.

The Institute for Environmental Management (KGI) organized and coordinated the assessment under the supervision of the Ministry for Environment and Regional Policy

NATO ASI Series, Partnership Sub-Series, 2. Environment – Vol. 1
Clean-up of Former Soviet Military Installations
Edited by R. C. Herndon et al.
© Springer-Verlag Berlin Heidelberg 1995

(KTM). The assessment was devised with contributions from the Environmental Inspectorates, National Parks, Nature Conservation Directorates and the Hungarian Army.

1.1 Characteristics of Environmental Damages

Forty percent of the damages are related to soil and groundwater pollution, 28% are related to fauna, flora and landscape, 18% was associated with accumulated and irregular dumping, and 14% are associated with miscellaneous causes. Several of the former Soviet Army sites required remediation. The types and magnitudes of the environmental damages are:

- Hydrocarbon (HC) pollution - soil and groundwater pollution by free phase, dissolved or bound kerosene, diesel oil and petrol at airports, fuel storage tanks and repair and maintenance shops.
- Heavy metal pollution - polluted soil containing copper, lead, cadmium, chromium, and arsenic at shooting grounds and vehicle repair shops.
- Wastewater and sewage sludge - soil and groundwater pollution at low efficiency sewage treatment plants and from ground disposal of wastewater.
- Other wastes - wastes (partially covered with earth) disposed of at each military base including inert construction and demolition debris (70-80%), municipal wastes (20-30%), and hazardous wastes (1-5%).

HC products contaminated about 2.7-3.0 million m^3 of soil and 1.0-1.2 million m^3 of subsurface water resources. Approximately 5,500-6,000 m^3 of free phase HC products were detected floating on top of the groundwater. About 200,000-220,000 m^3 of various solid wastes were also found, partly spread on the surface, and partly buried (not including construction and demolition debris).

1.2 Establishment of Environmental Damage Remediation

Based on results from the assessment, the KTM submitted a plan to the Hungarian Government concerning prioritization of remediation and cost estimates (see Table 1). Due to restricted funding, the remediation was planned to take place over a long period, from 1991-2010. One primary goal defined in this plan was to immobilize pollutants as soon as possible. The plan outlined tasks, remediation priorities, and cost estimates of the remedial efforts to be carried out. Elimination of damages was classified into three groups:

- a) Damages requiring immediate intervention - At 20 of the investigated sites, the migration of harmful material (from waste improperly disposed) in polluted soils or groundwater posed significant risks to the drinking water resources. These wastes had to be contained immediately. Containment costs were estimated to be 930 million Fts.

Table 1. Flowchart of the Hungarian Site Remediation Process

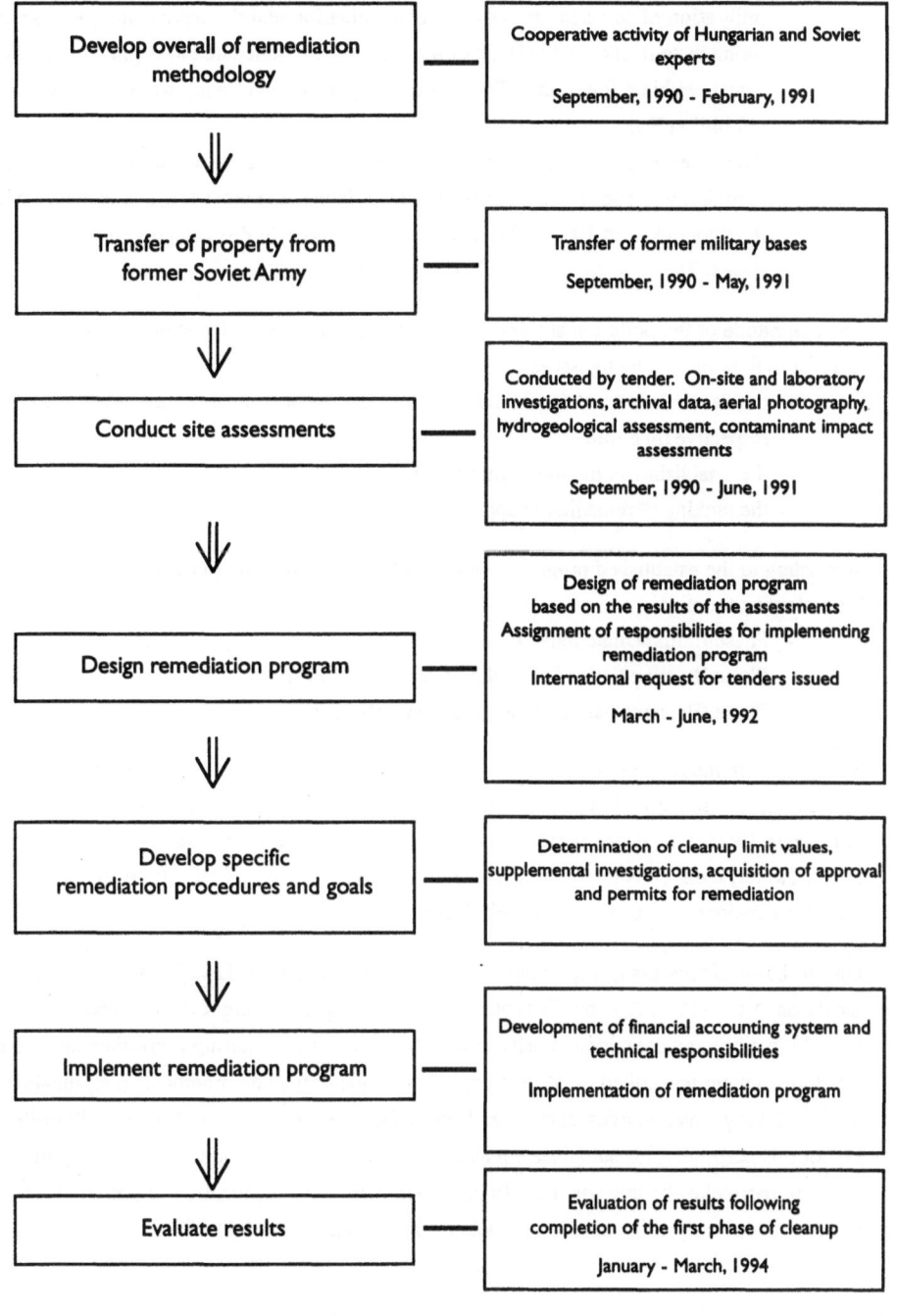

b) Damages requiring intervention in the short-term - damages at 82 military bases (including the 20 from the first category) were determined to have slower migration of soil and groundwater contamination and the wastes did not pose an immediate danger. This stage of the cleanup was scheduled to begin in 1993 and finish within a few years. The cost of this part of the cleanup was estimated to be 30 billion Fts.

c) Damages requiring intervention in the long-term - Remediation of damages pertaining to construction debris, flora and fauna, and other low risk damages will be undertaken in 1995. The efforts at the 77 bases in this category may take 5-15 years. The expected costs could reach 20 billion Fts.

The magnitude of the pollution at these sites, as well as the financial situation of the country, brought to light the following issues:

- the determination of applicable and practical decontamination processes and their respective time frames;
- the establishment of environmental priorities; and
- the ranking of remediation approaches.

According to the established program, the three phases of rehabilitation will take place using the following timeline:

- Phase I: immediate remediation (1992-1993);
- Phase II: remediation in the short term (1994-1995);
- Phase III: remediation in the long term (1996-2010).

The 20 sites in the first phase include 7 airports, 1 heliport, 3 fuel bases and 9 barracks. Eight of these sites were defined as posing significant threats to populated areas. The remaining 12 sites were located in water protection zones of high-priority subsurface water resources (groundwater, bank-filtered water, uncovered karst topography). Mobility of the pollution was also a major consideration for these 20 sites.

On the basis of this plan,. the Government in 1991, passed the Law XXX decreed in its Decision No. 3461/1991 on October 24, 1991, implementing the localization (i.e., immobilization) and containment tasks at the 20 Phase I sites requiring immediate attention. The Government prescribed the urgent protection of the critical environments (i.e., inhabited areas, drinking water sources, surface and subsurface waters, etc.) and allocated 930 million Fts for this purpose. This sum does not allow for complete remediation of the highly polluted areas, in particular the airports; therefore, immobilization or localization of pollution was a primary goal, along with extraction of mobile contaminants, especially HC.

The government determined that the cleanup of the environmental damages are the responsibility of the prevailing owner of these sites. This obligation must be stated in writing when these sites are transferred to new owners as either a gift or a sale. The contracts of sale or transfer are to be presented to the KTM for review before they are finalized.

2. Preparations for Remediation

The implementation of the Governmental Decision was addressed, and after revision as an Action Program, it was passed by the KTM on March 30, 1992.

2.1 The Action Program of the Ministry for Environment and Regional Policy

The first phase of remediation according to the Action Program was to contain the migration of environmental pollution at the 20 priority sites. This included thorough remediation of pollution at some of these sites. The KGI had been assigned to carry out this program and to conclude the contracts. Remediation efforts at the 20 bases were implemented as follows:

- Cleanup began at Tab, Veszprém and Székesfehérvár in 1990;
- Remediation is planned for Tököl, Sármellék and Szombathely, and will be carried out by the KGI;
- Institutions will be assigned for the cleanup of the following sites by a tender: Kiskunlacháza, Kalocsa, Esztergom, Lovasberény, Kunmadaras, Hajmáskér, Pétfürdö, Mezökövesd, Györ, Debrecen, Vác-Máriaudvar, Dombóvár-Kaposszekcsö, Szentendre, and Komárom.

The Action Program defined the tasks, the financial schedules, the forms of financial and technical control, the recording, the methods of payment and the responsibilities. A committee was set up to oversee execution of the Action Program. Represented on the committee were the KTM, the National Authority for Environment, the Environmental Inspectorates and the KGI. The main tasks of this committee were to determine the conditions of the tender and to identify the companies to execute the remediation.

2.2 Tendering

Pre-qualified firms were invited to take part in a tender. For the remediation of the 14 sites, separate tenders were published with fixed total amounts. Ninety-six firms responded to the tender, 56 of which were Hungarian.

Ten firms obtained authority to carry out the remediation. The majority of these worked with both domestic and foreign subcontractors. It is expected that participation of foreign firms will not only lend financial support, but will also contribute high-level technical knowledge and hands-on experience to the remediation activities. In only one case did a foreign firm (Danish) contribute financial support for remediation. The majority of the foreign bids only covered assessment activities.

Some Hungarian firms chose foreign firms (e.g., BGT, Germany; SGI, Switzerland) as partners. The foreign firms provided mostly planning, handling of technology, and the guaranty of equipment. Some foreign firms, such as SAKOSTA (Austria) or Carl Bro Group (Denmark), formed a joint venture with Hungarian firms to perform tasks.

The Carl Bro Group obtained the cleanup of the fuel bases in Pétfürdö, resulting in a cost reduction through the use of equipment bought with financial assistance from the Danish government. The cleanup of the airport at Tököl was aided by Danish consulting and equipment. The total support at both of these bases totaled 1 million Danish crowns.

Work on the feasibility study for cleanup at the airport at Sármellék was performed by the American firm CH2M HILL, sponsored by the U.S. Trade Development Program.

3. Realization of the Cleanup

Cleanup work started gradually due to the financial situation of Hungary. In the initial phase, the mobile pollution was localized and drawn off. Free-phase HC located in high-priority zones were drawn off, and at some sites highly HC-contaminated soils were cleaned.

For the remediation which began in 1992, water resource protection was a priority, especially around the bases. Considering this, the environmental authorities prescribed limit values for remediation for:
- dissolved HC concentrations in subsurface waters; and
- fixed HC concentrations in soil.

Execution of the remediation

After firms were assigned to the remediation projects, execution plans were made based on supplementary investigations. These plans addressed the following issues: specified the character, magnitude, and extent of the pollution; specified the various alternatives for cleanup; and the specified methods and technologies for remediation selected on the basis of technical applicability, cleanup regulations and costs. The plans were eventually approved by the appropriate authorities. Remediation was supervised using a technical advisor.

Principle features of the rehabilitation

Cleanup activities included the localization of HC pollution at airports, fuel storage areas, and bases, along with the extraction of polluted groundwater containing free-phase and dissolved HC. The extracted groundwater was purified primarily by air stripping; in some cases, filtration with activated carbon was used. Polluted soil was cleaned through soil aeration and bioremediation.

The applied technologies and equipment

The extraction of free-phase HC pollution and polluted groundwater was executed depending on the degree and location of the pollution and on the soil-type data. Structures used for extraction included wells, drains, and open trenches. Extraction technologies included:

1. pumping of groundwater and HC pollution with the same pump in a combined system;

2. extraction of groundwater and HC pollution with special pumps in separate systems (e.g., Scavenger and CEE); and

3. extraction of the HC pollution only (via skimming).

Treatment technologies included phase separation, air stripping, and filtration using activated carbon.

The treated groundwater was disposed of either by a sprinkling method or via release into wells and trenches promoting infiltration into the soil or, in some cases, by release to surface waters. The elimination of HC products extracted from less polluted soils was through incineration, with secondary recovery and re-use of the generated heat.

Other applied technologies included soil aeration (with air injection and vacuum) for removal of volatile HC pollutants, and bioremediation for removal of fixed HC pollutants in soil.

4. Results of the Remediation

As of December 31, 1993, the prescribed tasks of the Governmental Decision had been completed to the following extent (see Table 2):

- At 8 sites, localization of pollution as well as complete remediation were accomplished in accordance with the Action Program. These sites are: the Hunyadi barracks at Tab, the Kossuth barracks at Veszprém, the heliport of Székesfehérvár, the barracks for armored car troops at Esztergom, the barracks at Lovasberény, the barracks and fuel bases at Győr, the barracks at Szombathely, and the barracks at Dombóvár-Kaposszekcsö.

Table 2. Results of Remediation

Site	HC extracted (liters)	Groundwater treated (m^3)	Soil treated (m^3)	Waste removed (m^3)
1/100 Tab	-	-	9,617	-
2/14 Veszprém	-	-	43,080	-
3/19 Székesfehérvár	-	-	16,234	-
4/39 Tököl	223,853	279,115	-	-
5/60,61 Sármellék	146,600	37,600	-	-
6/58 Szombathely	7,900	11,500	1,034	1
7/38 Kiskunlacháza	170,000	180,000	2,134	-
8/76 Kalocsa	520,000	100,000	6,068	-
9/89 Esztergom	below limit value	-	5,200	8
10/2 Lovasberény	below limit value	-	2,000	-
11/88 Kunmadaras	80,000	30,000	-	-
11/88 Kunmadaras (phase II)	35,900	33,370	-	-
12/27 Hajmáskér	below limit value	-	18,000	2,000
13/26 Pétfürdö	45,000	240	-	-
13/26 Pétfürdö (phase II)	16,000	70	-	-
14/98 Mezókövesd	15,000	20,808	9,740	-
15/90 Gyór	below limit value	30	7,120	260
16/108 Debrecen	51,000	50,000	2,235	-
17/28 Vác-Máriaudvar	below limit value	13,500	2,940	-
18/99,102 Dombóvár Kaposszekcsö	26,000	-	3,240	456
19/93 Komárom	2,000	7,500	375	1
19/93 Komárom (phase II)	7,382	2,224	-	-
20/31/41 Szentendre	below limit value	-	300	160
20/31/41 Szentendre (phase II)	below limit value	250	66,000	30
Totals	1,346,735	766,207	195,317	2,916

- At the other 12 sites, environmental pollution, and consequently environmental risks, have been significantly reduced. These sites are: the airports at Kunmadaras, Tököl, Sármellék, Kiskunlacháza, Kalocsa, Debrecen and Mezökövesd; the fuel bases at Pétfürdö; the barracks at Hajméskér; the barracks and fuel bases at Vác-Máriaudvar; the barracks and fort of Komárom; and the barracks of Szentendre.

- 1.3 million liters of HC products (kerosene, diesel oil, and petrol) were extracted and treated; and 586,000 cubic meters of polluted groundwater were removed and treated. Finally, 195,000 cubic meters of soil and 2,900 cubic meters of other wastes were decontaminated and managed under well-controlled conditions.

- Monitoring systems were developed which allow detection of future environmental contamination.

- Foreign financial assistance was less than expected; nevertheless, up-to-date technologies were implemented.

- An expert advisory group was formed to assess the environmental pollution, to draw up plans to remediate the environmental damage.

- The process of remediation was significantly promoted by the support and contributions of the ministries and organizations (e.g., Treasury Estate Managing Corporation, KVSZ; National General Health and Medical Officer Service, ÁNTSZ) and of the Environmental Inspectorates concerned with the fulfillment of the Action Program. The KGI, the leading institution in damage assessment, was charged with the task of general organization.

5. Future Tasks

Future (Phase II) tasks were determined in 1993. The program of activities for 1994 was specified in the Governmental Decision No. 3360/1993 (October). Its implementation will permit continuation of Phase I activities at 12 sites. In addition, based on a comprehensive assessment of the 171 former Soviet military bases, a remediation program will be worked out for the next five years.

Short review of the program

The program took into account recommendations and comments made by both the KVSZ and the environmental authorities of the local governments during the interdepartmental agreement negotiations. Based on a recommendation from the KTM, future tasks can be placed into two groups, to be undertaken simultaneously:

1. Remediation efforts at the remaining 12 sites in Phase 1 should continue according to a risk analysis performed by the state after the localization work is completed. The goal of this activity is to eliminate the environmental impacts of the contaminated soil, groundwater and wastes found in large amounts, for the

purpose of preserving the endangered water resources and settlements. If these tasks are not continued (and completed in a short time) the spread of pollution will begin again from the residual soil pollution. This would jeopardize the results of the completed work and increase the overall costs of required remediation significantly.

2. At the other 151 investigated sites, it is required - according to Governmental Decision No. 3360/1993 (October) - to select those military installations where:

   - based on their condition in 1993, the environmental damage is not in need of remediation; therefore, a buyer of one of these properties would not be charged with cleanup;

   - the degree and nature of the environmental risks justify the start of remediation using budgeted funds; and

   - only monitoring is necessary at the present time, since natural degradation processes are sufficient to possibly make the property usable within a few years.

This task will include: a review of the environmental assessment performed in 1991, the definition of environmental impacts, the prioritization of activities, and the elaboration of the 5-year program for remediation. To attain these goals, it is advisable to perform the tasks in 4 distinct work phases:

   a) evaluate damage assessment with respect to environmental risk;

   b) perform additional measurements and investigations;

   c) conduct environmental risk analysis for the 5-year program (1994-1998) for remediation; and

   d) complete and disseminate the documentation for the tenders.

Implementation of similar technical tasks at other former Soviet military bases in Hungary can be expected as a result of the success of this project.

DEMONSTRATING INNOVATIVE TECHNOLOGIES AT

ABANDONED FORMER SOVIET MILITARY BASES

Erno Kiss
Ministry of Environment and Regional Policy
Fó u.44
H-1011 Budapest
Hungary

Soviet troops withdrawing from the Hungarian Republic left behind 171 garrisons, 340 settlements, and 6,000 buildings occupying 46,000 hectares of land. In some regions, the withdrawing troops left situations behind which required immediate action. Environmental damages at the abandoned military bases are in the following general areas:

- 40% of the damage was to soil and groundwater supplies;
- 28% of the damage was to fauna, flora and landscape;
- 18% of the damage was associated with improper waste disposal; and
- 14% of the damage was associated with miscellaneous causes.

The most characteristic environmental damages are:

- hydrocarbon (HC) contamination (including jet fuel, fuel oils and benzene) in dissolved and bound form in soil and groundwater in the vicinity of airports, fuel storage areas and vehicle repair shops;
- heavy metal contamination (including Cu, Pb, Cd, Cr, As) in the vicinity of shooting ranges and vehicle repair shops; and
- waste water sludge contamination as a result of either low efficiency or a lack of waste water treatment systems.

The general categories of waste found at these sites are:

- inert construction and demolition wastes (70-80%);
- municipal solid wastes (20-30%); and
- hazardous wastes (1-5%).

NATO ASI Series, Partnership Sub-Series, 2. Environment – Vol. 1
Clean-up of Former Soviet Military Installations
Edited by R. C. Herndon et al.
© Springer-Verlag Berlin Heidelberg 1995

HC contaminated 2.7-3.0 million m^3 of soil and 1.0-1.2 million m^3 of groundwater. At the time of the assessment, about 5,500-6,500 m^3 of free-phase HC were found floating on the surface of groundwater. At these former military bases, 200,000-220,000 m^3 of varied solid wastes were found on the surface and underground.

The HC contamination required immediate environmental remediation. At 20 bases the priority was to mitigate the migration of the HC contamination in order to protect the drinking water supplies. The free-phase HC contamination was pumped out and, in some areas, the HC contaminated soil was also cleaned up.

For these remediation efforts, three levels of activity are possible:
- localization (i.e., control the spread or migration of contamination);
- partial remediation (e.g., pumping of the oil plume); and
- complete remediation.

Figures 1 and 2 illustrate the technologies which could be considered for these remediation activities.

The following is information relating to the remediation technologies applied at the former military bases of Sármellék, Kunmadaras, Vác-Máriaudvar and Tököl.

I. Remediation of Sármellék

The Sármellék Airfield (SAF) was the largest Soviet airport in western Hungary having an area of 3.85 km^2. The SAF is southwest of the village of Sármellék and Zalavár. There are potable water supply systems and sewers in parts of the area. A wetlands area called the Kisbalaton lies between Sármellék and Lake Balaton, a fresh water lake, about 6.5 kilometers east of the SAF. Lake Balaton is used extensively for recreation. The Kisbalaton area drains into Lake Balaton (Figure 3).

The SAF site, as shown in Figure 4, is about 3.5 km long and 1.3 km wide. The major feature of the airfield is a 2.5 km runway. Located on the northwest and the east sides of the site are 42 aircraft hangars, former ammunition magazines, and 12 earth-covered bunkers. The airfield was constructed in 1952 to serve the Hungarian Army. Beginning in 1962 it was used by the Soviet troops as a military airfield.

The airfield was contaminated with HC which were detected at four separate sites (A-D) (Figure 5). At these sites, the HC, in particular kerosene and benzene, was partly bound in the soil, and partly floating on the surface of the groundwater.

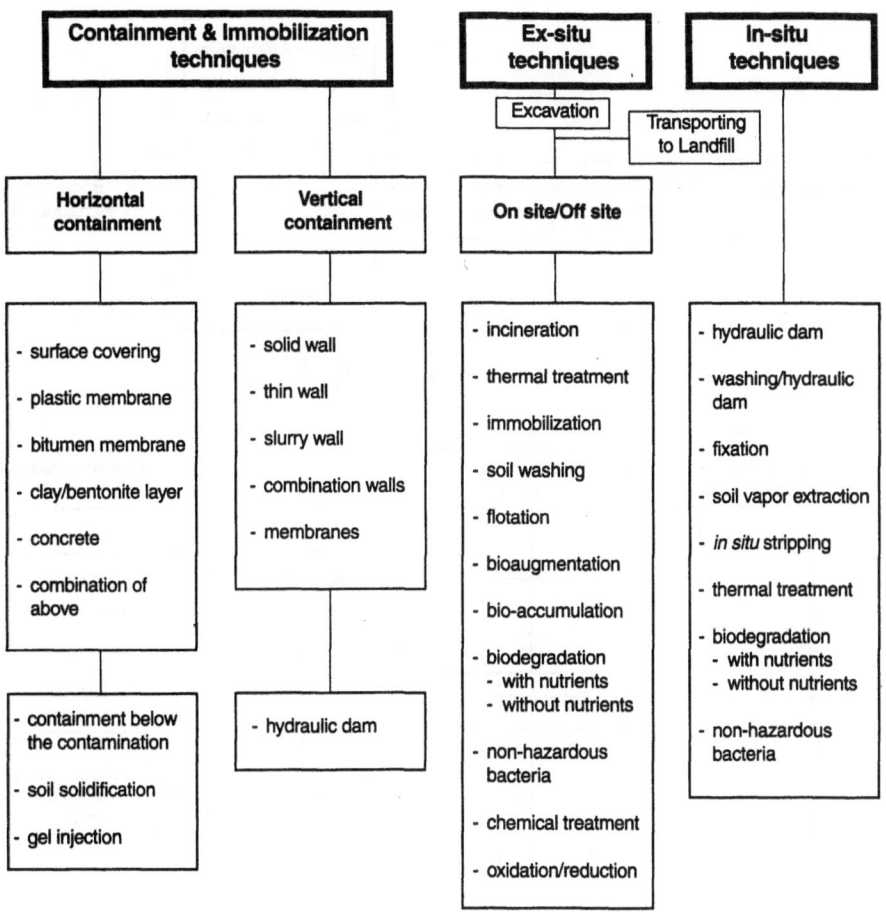

Figure 1. Review of technologies for remediation of oil contamination

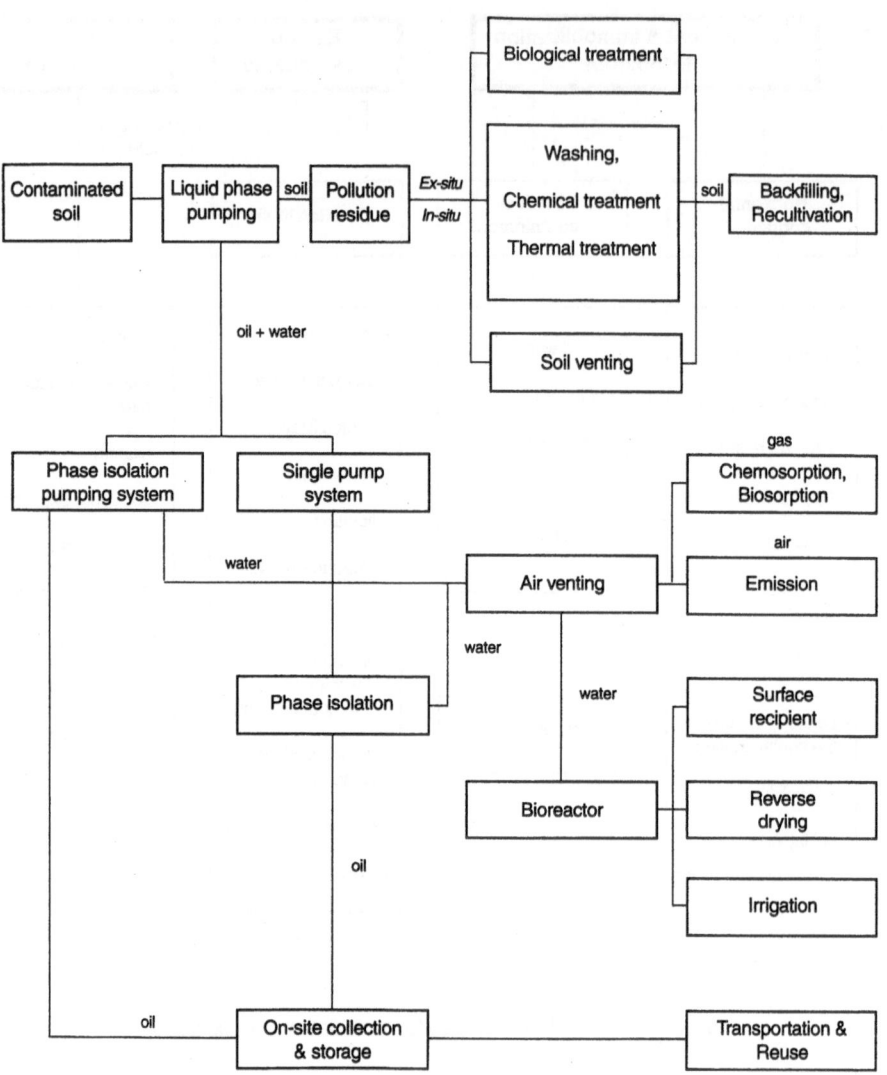

Figure 2. Remediation of oil contamination

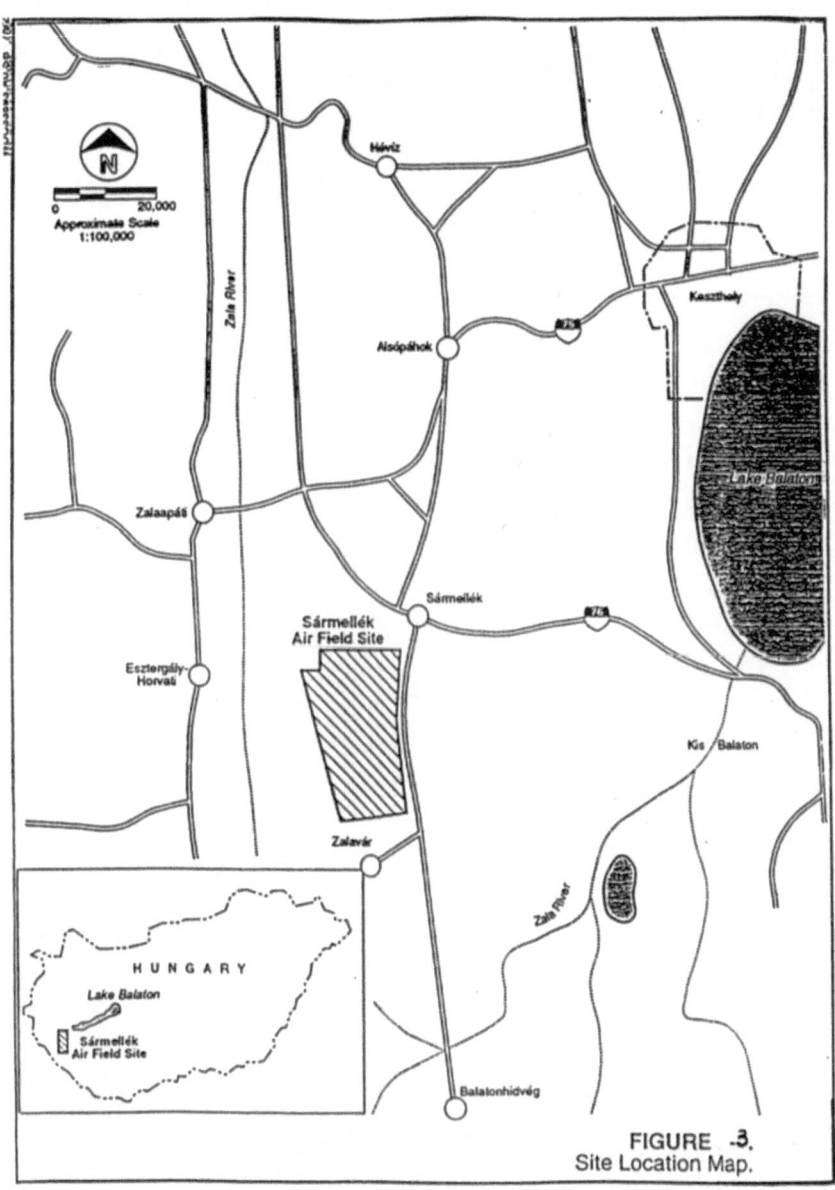

FIGURE -3.
Site Location Map.

FIGURE 4.
Base Map for the SAF Site.

FIGURE 5.
Waste Site Locations.

I-1. Remedial Action

From 1992-1994, the following remedial actions were carried out:

- localization (or control) of the kerosene plume migrating toward the village of Sármellék;

- implementation of an alternate water supply system, where feasible, to villages with affected wells;

- remediation of contaminants affecting nearby residences;

- localization (or control) of migrating kerosene near the large fuel storage tank area (area D); and

- removal (via pumping) of the HC layers in areas A, B, and C.

I-2. Remedial System

In area D, infiltration systems and a slurry wall were implemented. Trenches which were 1,170 m long, 6 m wide and 3 m deep were excavated. The three deep infiltration systems were configured as follows: length of 1,173 m; depths of 6 m and 4 m; built-in tube length was 1,235 m; infiltration drain gravel was 440 cubic meters. The work was carried out by the Hölscher Wasserbau company (Figure 6).

An impermeable slurry wall, 40 m long and 6 m in depth, was also installed. Three recovery wells were built for the removal of the fluid, each having an inner diameter of 300 mm and set to a depth of between 16.8-20.1 m. The length of the filter pack was 1.5 m (Figure 7). One recovery well was placed in each of the smaller areas (A, B and C) having an inner diameter of 300 mm and its bottom depth of 21.4 m. The length of the filter pack was 10 m.

Similar methods were installed in two different areas (Figures 8 and 9). Small Diameter Filter Scavenger oil pumps and plunger pumps were used in the recovery wells for pumping of the free-phase floating HC. The larger materials were removed from the HC, and the HC was then separated and transported. In areas A and B, mixed phase HC was pumped out with the same pump. The groundwater was removed to a two-basin treatment system and the HC was skimmed, and then air stripping was carried out.

Twelve wells of the community were contaminated with HC. These were cleaned by removing the HC-contaminated water via pumping and refilling each well three times. The wells were then washed, pumped out again and chlorinated.

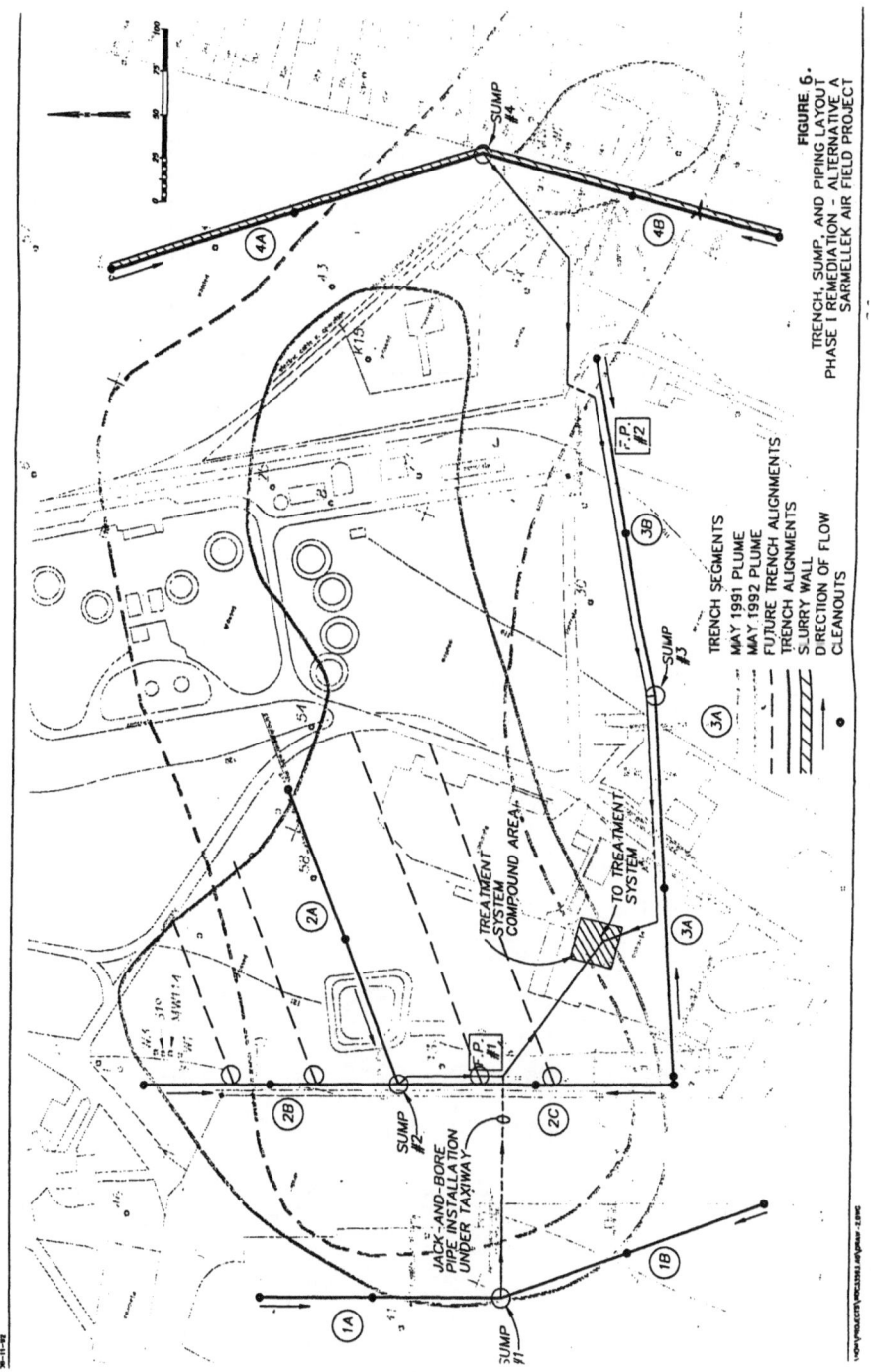

FIGURE 6.
TRENCH, SUMP, AND PIPING LAYOUT
PHASE I REMEDIATION - ALTERNATIVE A
SARMELLEK AIR FIELD PROJECT

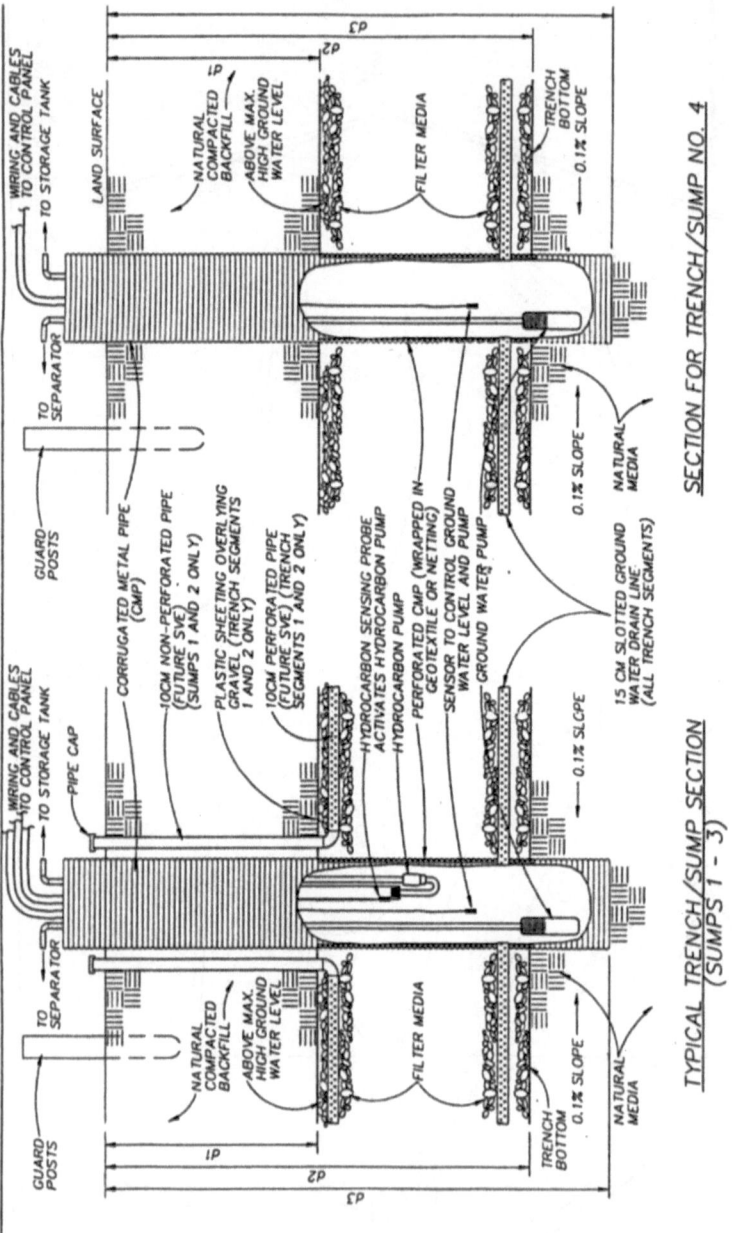

TYPICAL TRENCH/SUMP SECTION
(SUMPS 1 - 3)

SECTION FOR TRENCH/SUMP NO. 4

FIGURE 7.
INTERCEPTOR TRENCH, SUMP, AND PUMP DETAILS -
PHASE I REMEDIATION - ALTERNATIVE A
SARMELLÉK AIR FIELD PROJECT

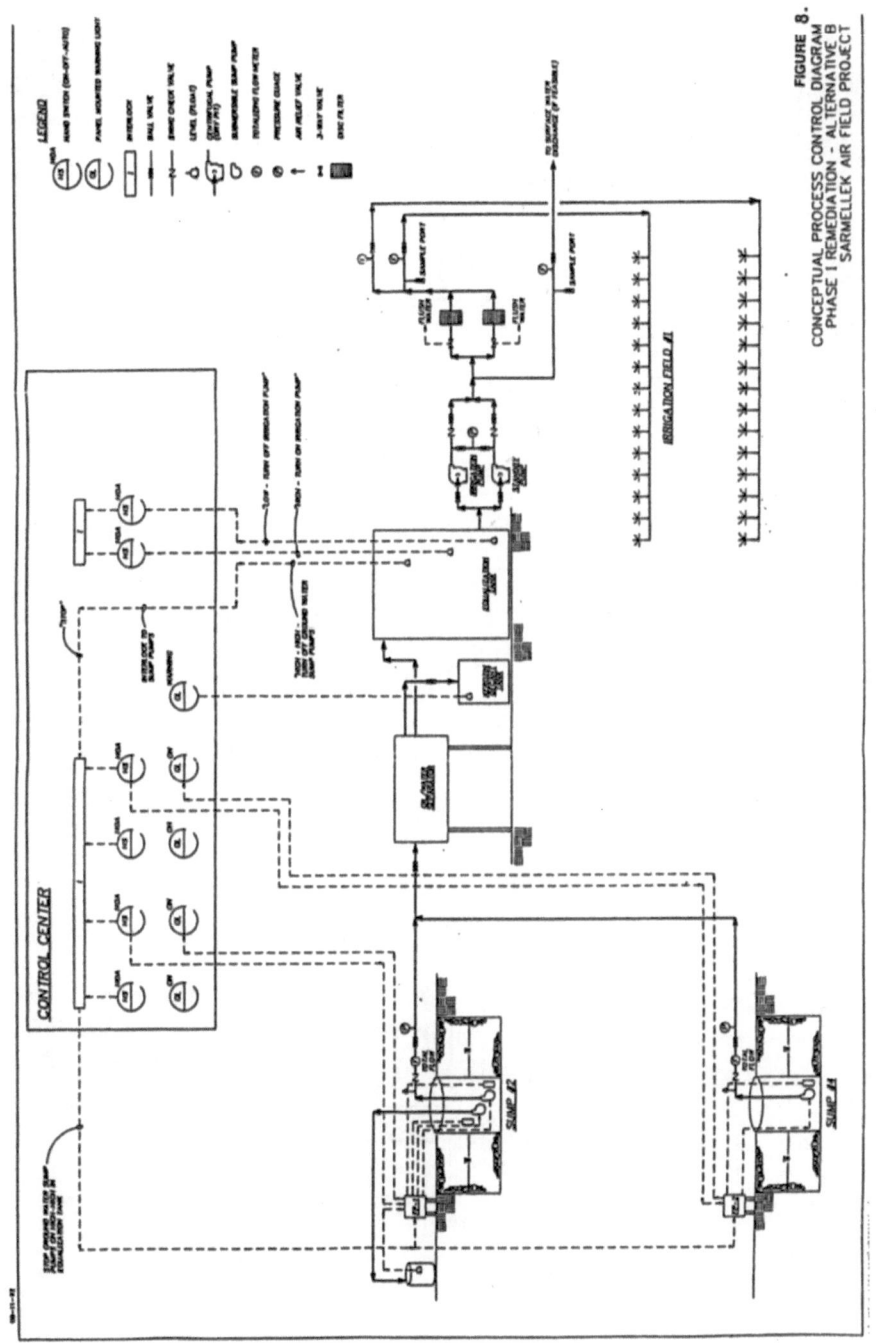

FIGURE 8.
CONCEPTUAL PROCESS CONTROL DIAGRAM
PHASE I REMEDIATION - ALTERNATIVE B
SARMELLEK AIR FIELD PROJECT

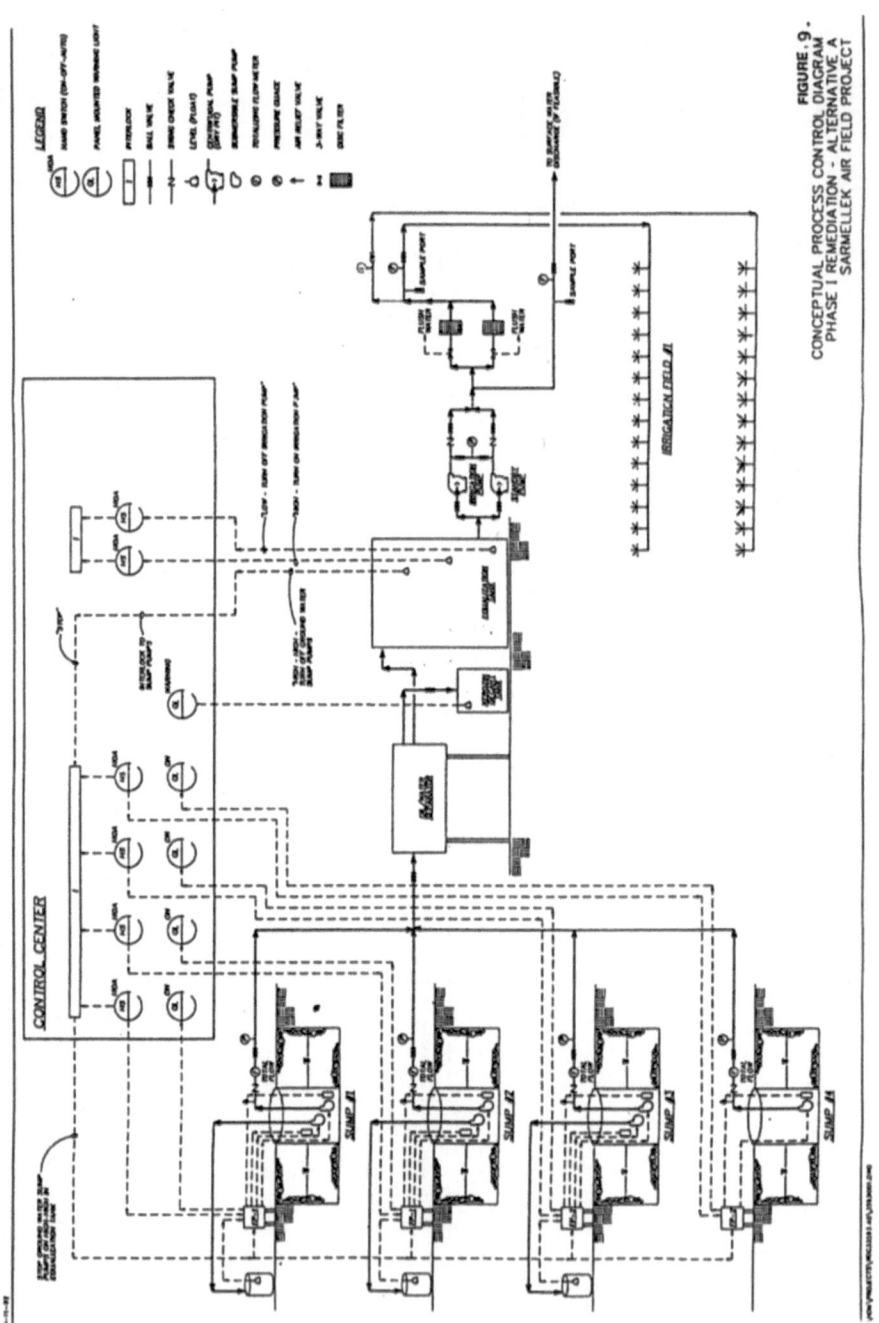

FIGURE 9 -
CONCEPTUAL PROCESS CONTROL DIAGRAM
PHASE I REMEDIATION - ALTERNATIVE A
SARMELLEK AIR FIELD PROJECT

II. Remediation of the Kunmadaras Airfield Site

This airfield site is located in the town of Kunmadaras and is about 756 hectares in size; its surface is relatively flat and there are large wet areas (e.g., wetlands) present. In the boring samples taken at 0.5-3 m, a thin, dark brown/black humic clay layer was found. Under the clay layer, debris consisting of river stones and other materials associated with flooded areas were found.

Underground HC pollution was found at 8 separated sites. The volume of the free-phase HC was approximately 300 m^3. The HC concentration bound in the soil was 0.1-2.2 kg/metric ton.

II-1. Localization of free-phase HC contamination

For the localization of floating HC, a system was designed which can simultaneously pump out the floating HC as well as the groundwater. For this purpose a trench was built to a depth of 1 m below the minimum groundwater level.

The HC contamination was surrounded by a trench, and it was immobilized using a dual-use inverted, gathering drain system. At the start of the remediation, existing wells were used and clean water was fed in by gravitational flow through the inverted drain which was laid down at a 1 m depth. Through the use of the parallel drains (separated by 16-20 m), the area was flooded and the groundwater level was gradually raised (Figure 10).

By raising the groundwater level, the HC pollution was also raised and the groundwater gradually conveyed the HC floating above the groundwater level vertically upward until the HC reached the discharge level of the gathering drain. Then the HC-polluted water began to discharge to the drain. The morphology of the subsurface started the slow migration of the floating HC towards the discharge drain. As a result of the poor infiltration conditions, even small amounts of water were enough to start the slow, stationary flow, which together with the water morphology, facilitated the mobilization of the HC.

The discharge drains were connected to the trenches and the contaminated water was led by gravitational flow to the gathering tube of the trench. The gathering tube of the trench conveyed the contaminated water to a collecting well located 150 - 200 m from the trench.

KUNMADARAS

SCHEME OF THE IDEAL MODE

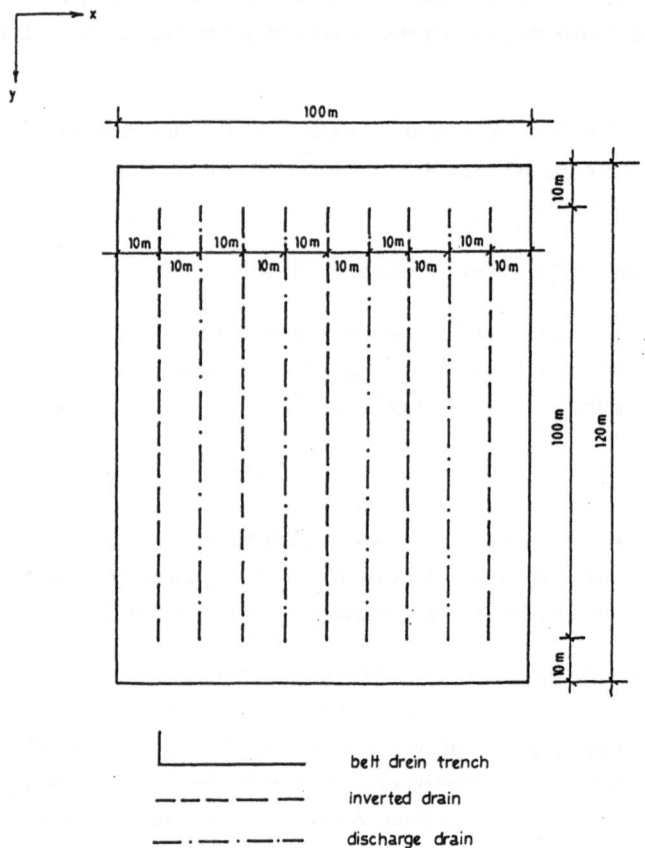

belt drein trench

inverted drain

discharge drain

Figure 10

III. Soil Cleanup at Vác-Máriaudvar and Tököl

Soil cleanup was carried out at 8 of the 20 most endangered bases using technologies which were applied both *ex-situ* and *in-situ*.

III-1. *Ex-situ* applied technologies

Technologies were applied *ex-situ* at the Vác-Máriaudvar base. Here approximately 2,940 m^3 of soil were excavated. The soil was contaminated with diesel oil and motor oil at concentrations above 5 g kg^{-1}. After temporary storage, the sill-stack method was applied to degrade the HC contamination. The contamination was 10-40 g kg^{-1}. The soil was placed on a double plastic foil; cattle manure, along with nitrogen and phosphorous fertilizers were added and the combination was mixed and homogenized. A starter bacterial culture ("Biomix Oil") and water were also added. Five stacks were prepared of about 550-600 m^3 each. They were covered with black plastic foil and vacuumed with ventilators.

Land farming method - A "starter" was added to the soil mixture and then spread over a haybed of 20 cm thickness on a surface of 500 m^2. The thickness of the soil was 0.5 m. The surface of the soil was covered by various plants; first rye grass and then *Papillionaceae*.

At the József Attila storage base, which is one of the bases of Gyor, a "bioreactor" was established, to which the HC contaminated soil was transported from the neighboring bases of Komárom and Gyor. The soil contamination was 4-5,000 mg kg^{-1}. The bioreactor consists of a basin, the bottom of which is soil and which is isolated by plastic foil. Sawdust and hay were spread over the foil. The sawdust and hay are covered with contaminated soil 70 cm in thickness. The treatment is the same as in the previous technology, and lasts for 3-3.5 months for the remediation process. During this time the HC is biodegraded 10- to 20-fold.

III-2. *In-situ* applied technologies

III-2.1 The following is information on the soil bioventing system which was applied at the Vác-Máriaudvar base. This system consists of air venting wells, water separators and an activated carbon filter. This method allows the aromatic HC content of the gas to be kept below the maximum concentration limit (see Figure 11).

III-2.2 Bioventing plant for Tököl airbase - Bioventing is a commonly applied method to aerate unsaturated soil. It differs from the soil-venting techniques (vacuum extraction, stripping) in the rate of gas exchange. The object of soil venting is to strip the compounds from the soil and is, thus, well suited for treatment of volatile compounds. When applying

OPERATING SCHEME OF THE SOIL BIOVENTING
SYSTEM

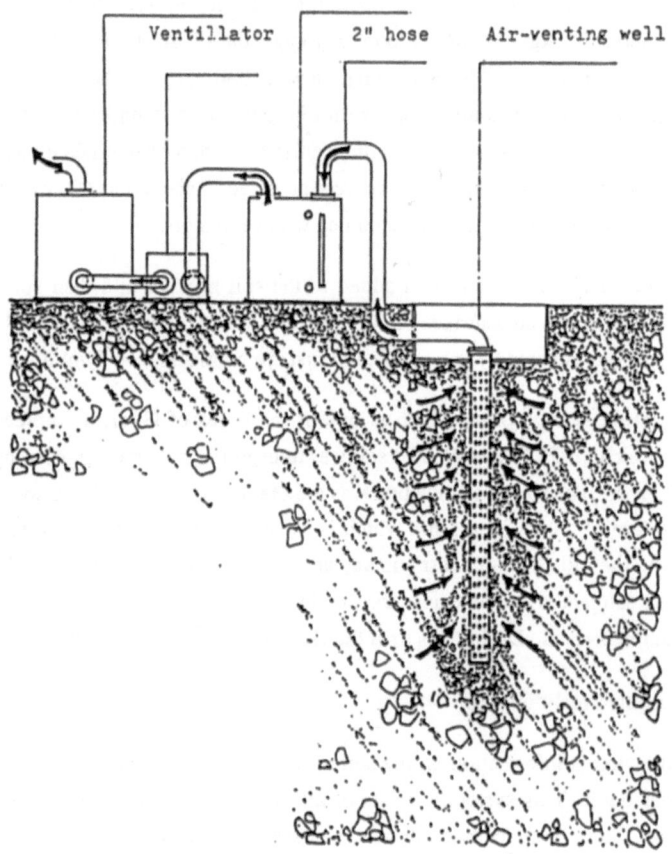

Figure 11.

soil-venting techniques, treatment of the trapped gases is generally required. In the present case, jet fuel is the contaminant. It consists mainly of low-molecular weight aromatic and aliphatic hydrocarbons that are biodegradable. The contaminated area at Tököl airbase consists of sand and gravel with high permeability. The site, therefore, offers favorable conditions for bioventing.

A drawing of the pilot-scale bioventing plant is shown in Figure 12. The bioventing plant consisted of an extraction well from which air was extracted from the soil. It had a depth of 4 m and contained a filter with a diameter of 100 mm and a 2 m screened section. The extraction well was tightly closed by a cover connected to a vacuum pump. To ensure a vertical stream of air through the contaminated soil, four passive injection wells with a 50 mm filter and a 2 m screened section were installed. Figure 13 shows the removal rates both for recovered groundwater and HC as a function of time.

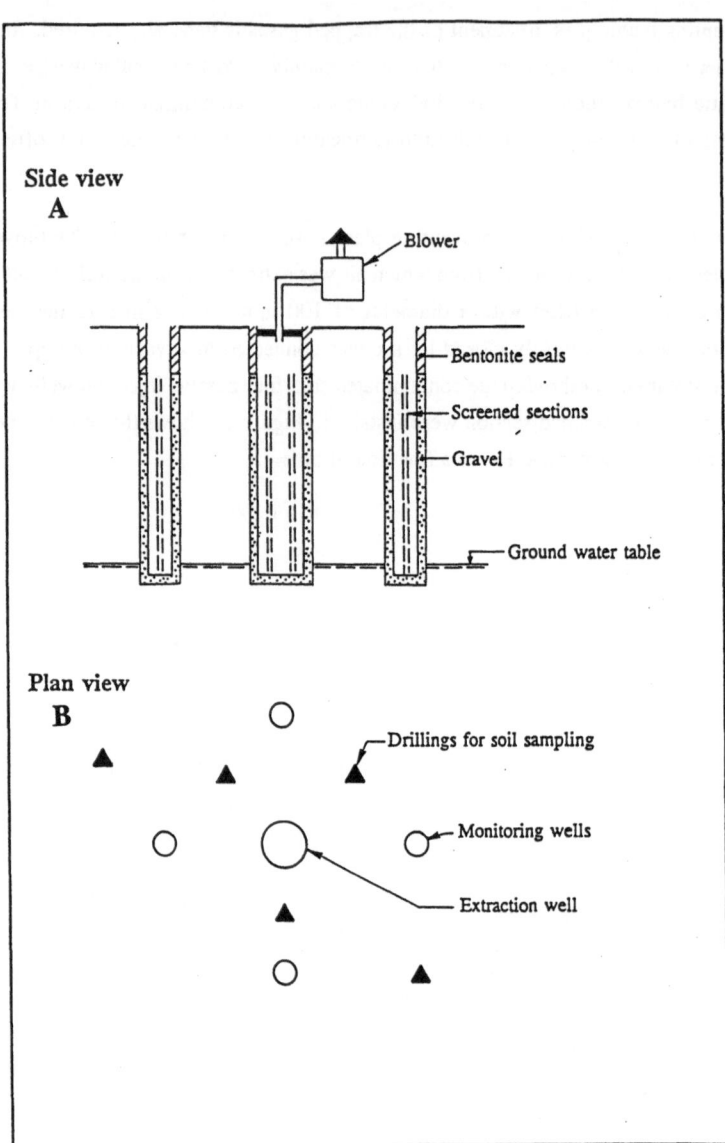

Figure 12 Diagram of the pilot scale bioventing plant. *A: Side view; B: Plane view.*

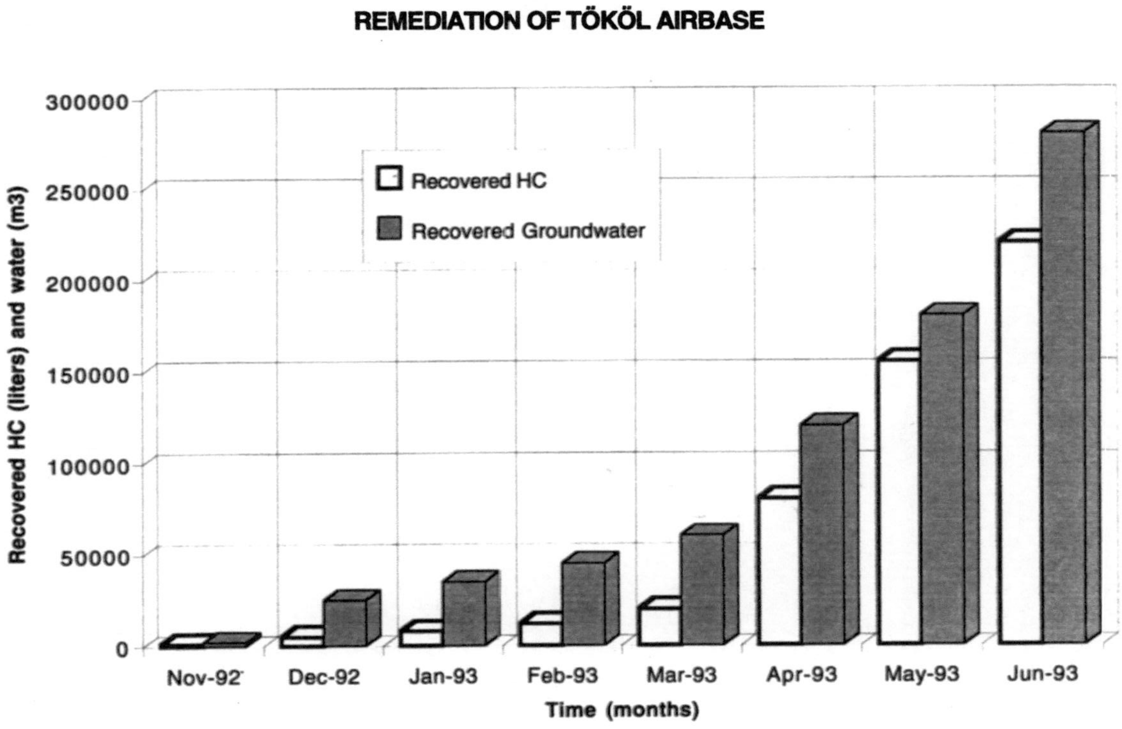
REMEDIATION OF TÖKÖL AIRBASE
Recovered HC
Recovered Groundwater
Recovered HC (liters) and water (m3)
300000
250000
200000
150000
100000
50000
0
Nov-92
Dec-92
Jan-93
Feb-93
Mar-93
Apr-93
May-93
Jun-93
Time (months)

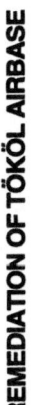

Figure 13

ENVIRONMENTAL PROBLEMS AT FORMER SOVIET MILITARY

BASES IN HUNGARY

Robert Reiniger and Zsolt Horvath
National Authority for the Environment
Fö utca 44-50
H-1011 Budapest
Hungary

As a result of the government agreement of March 19, 1990 concerning the withdrawal of the Soviet troops from Hungary, 100,000 soldiers, 25,000 weapons and more than 560,000 tons of war equipment were withdrawn between March 10, 1990 and June 10, 1991. In accordance with the agreement, an assessment of the environmental damage caused by the Soviet troops was undertaken. The assessment was managed by the Ministry for Environment and Regional Policy between September 21, 1990 and June 10, 1991.

Based on the results of the assessment, the Ministry issued 14 international tenders from which 10 firms were awarded contracts, mostly Hungarian firms, although some were with foreign firms. The work was carried out using methodologies accepted by both the Hungarian and Soviet governments. There were 171 garrisons, 340 settlements, 6,000 major buildings and 46,000 hectares of land to be surveyed. Sampling and damage assessments associated with the Soviet barracks and training grounds started on November 15, 1990, and were completed by May 9, 1991 (Figure 1).

The assessment estimated total environmental damages of 60.2 billion Hungarian forints (approximately $600 million (USD)).

The assessment revealed that the Soviet troops had caused considerably more environmental damage over a larger area than was initially estimated. Forty percent of the damage involved soil and groundwater contaminated with heavy metals and hydrocarbons. The most polluted are six military airports which are contaminated with jet fuels and fuel oils. Some of

NATO ASI Series, Partnership Sub-Series, 2. Environment – Vol. 1
Clean-up of Former Soviet Military Installations
Edited by R. C. Herndon et al.
© Springer-Verlag Berlin Heidelberg 1995

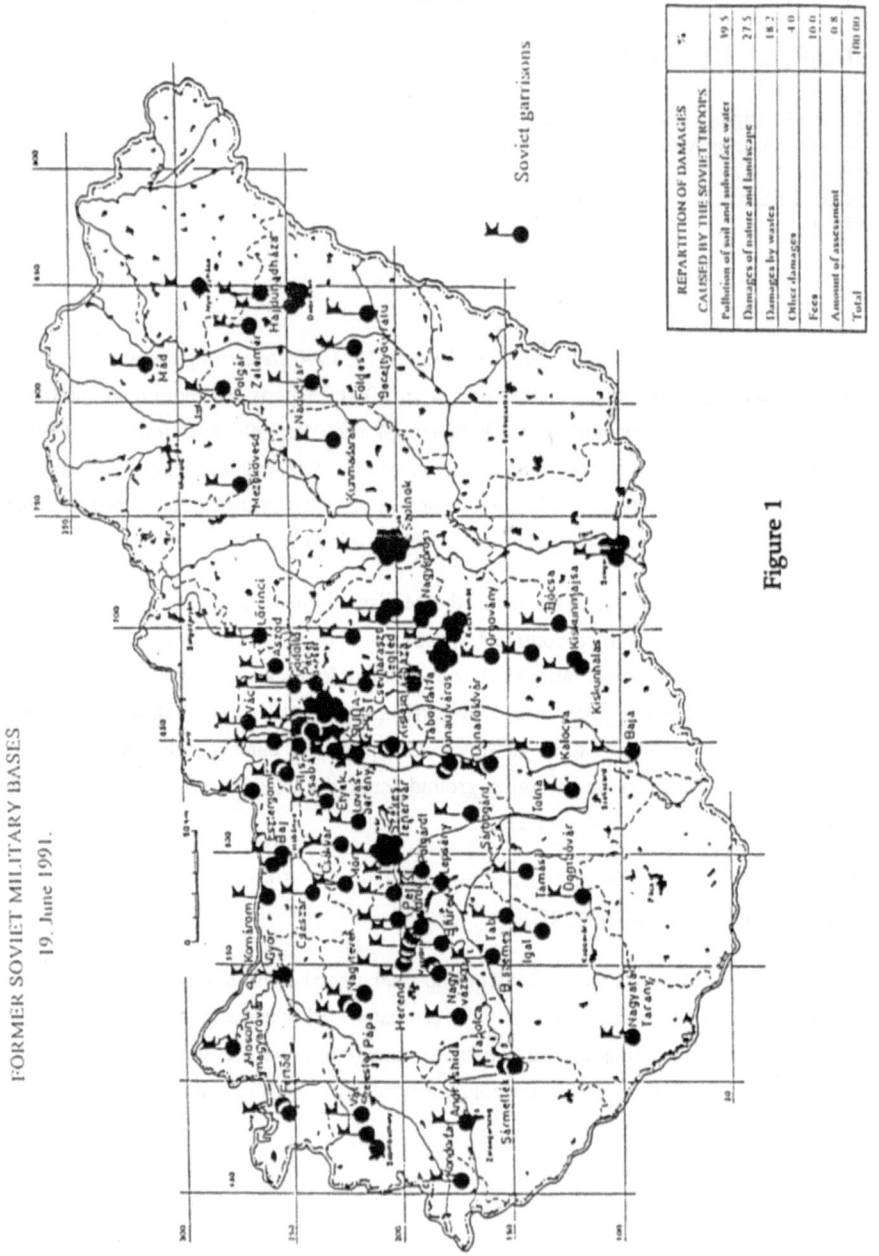

FORMER SOVIET MILITARY BASES

19. June 1991.

REPARTITION OF DAMAGES CAUSED BY THE SOVIET TROOPS	%
Pollution of soil and subsurface water	39.5
Damages of nature and landscape	27.5
Damages by wastes	18.2
Other damages	4.0
Fees	10.0
Amount of assessment	0.8
Total	100.00

Soviet garrisons

Figure 1

these sites required immediate action in order to prevent the spread of hydrocarbon contamination in groundwater. Delays would have resulted in increased remediation cost.

Hydrocarbon contamination affected 2.7-3.0 million m^3 of soil, and 1.0-1.2 million liters of groundwater. At the time of the assessment, about 5,500-6,500 m^3 of hydrocarbon product were found in free phase form on the surface of groundwater. There were 200,000-220,000 m^3 of waste in the vicinity of the barracks. About 70-80% of this waste was inert, 20-30% organic waste and 1-5% hazardous waste.

Because of limited financial resources and time, remediation work was started only at the 20 most polluted military bases in 1991. The goal of this initial remediation effort was to stop the spread or migration of the pollution (i.e., localization) by the end of 1991. There was a full remediation effort at 8 of the 20 sites, while localization of the pollution was pursued at the 12 remaining sites by the end of 1993 (Table 1). In 1994, planned remediation work at the 12 sites and the remaining 159 locations are being carried out.

Since 1991, a primary task was to sell the former military properties. Currently, the former Soviet military bases belong to various owners, such as the Government Trustee Office, the Ministry for Defense, local authorities and various foundations. To sell the properties, tenders needed to be developed and distributed. The major problems associated with selling the properties are the following:

-       the poor condition of the buildings and the public utilities; and

-       the contamination of the soil and groundwater.

Because part of the aim of the assessment completed in 1991 was to confirm the findings of the Soviet government concerning the extent and nature of site contamination, information about the cost of damages is not sufficiently detailed to use in auditing the actual site contamination. One of the aims of the 1993-94 assessments is to gather more data about the contamination level of the sites. Other objectives of the 1993-94 assessment include:

-       identifying those sites which do not have environmental problems, and which require no cleanup;

-       selecting those sites which must be cleaned up; and

-       determining those sites where only monitoring is required at present and cleanup measures can be determined after a period of observation.

Table 1

CLEAN-UP OF ENVIRONMENTAL DAMAGES, BEGUN IN 1992.

Nr.	Site	Environmental damages hydrocarbon pollution		Localization		
		groundwater	soil	Company	Sum (mFt)	Activity * (see below)
1.	Székesfehérvár-Helikopter heliport	+	11.060	Comco-Martech	26.003	6.
2.	Tab-Hunyadi János barrack	+	7.150	Comco-Martech	11.374	6.
3.	Veszprém-Kossuth Lajos barrack	+	69.510	Comco-Martech	54.054	6.
4.	Sármellék airport	+	300.792	KGI	66.281	1, 2, 3, 4.
5.	Szombathely-Huszár barrack	+ (6000 m³)	63.222	KGI	9.980	4, 5.
6.	Tököl airport	+ (1835 m³)	689.000	KGI	45.418	4.
7.	Dombóvár-Kaposszekcső	+	5.740	Comco-Martech	8.800	4, 6.
8.	Szentendre-Dózsa Gy. barrack	-	260	Hydrokliv	17.344	4, 5.
9.	Kunmadaras airport	+ (240,9 m³)	922.565	Alterra	87.522	2, 4.
10.	Kiskunlacháza airport	+ (241 m³)	900.000	Alterra	43.926	2, 4.
11.	Esztergom-Malinovszkij barrack	+	31.800	UNITEL	61.000	6.
12.	Győr-Frigyes laktanya barrack	+	760.000	DUKÖR	26.220	4, 6.
13.	Komárom-Árpád úti barrack	+	124.900	DUKÖR	8.800	4, 6.
14.	Mezőkövesd airport	+	312.000	ELGI	44.000	2, 4, 6.
15.	Debrecen-airport	+	825.000	ELGI	43.900	2, 4, 6.
16.	Kalocsa airport	+	287.450	EGI	44.750	4, 6.
17.	Lovasberény-barrack	-	49.000	DDKÖFE	26.400	6.
18.	Hajmáskér-barrack	+	107.855	Heves m. Települést.	43.500	5, 6.
19.	Vác-Máriaudvar fuel storage	+ (3500 m³)	1.131.513	BIOKÖR-Sakosta	26.000	4, 5, 6.
20.	Pétfürdő-fuel storages	+	630.000	Carl Bro Group	49.176	4.

* 1. Slurry wall.　2. Underground drain.　3. Dug-well decontamination.　4. Pumping-separating.　5. Aeration.　6. Biologic treatment

In order to realize these objectives, it is necessary to:

- evaluate the results of the assessments from the point of view of environmental risk;

- complete initial measurements and investigations;

- elaborate a 5-year program (1994-98) for the remediation of contaminated lands; and

- prepare an international tender for remediation activities.

Tököl Airfield

The following is information about the environmental problems at the Tököl airfield. The Tököl airfield can be found in the northwestern part of Csepel Island among the villages of Halásztelek-Szigethalom-Tököl (Figure 2). The western boundary of the site is 600 m from the Danube River. The airfield is about 1 km from Halásztelek and about 1.2 km from Szigethalom and Tököl.

The Halásztelek well field is about 0.6 km from the boundary of the airfield, and nearly the entire territory of the airfield is inside the hydrogeological protective area of the well field. The well field belongs to the Capital Waterworks and provides about 5% of its yield. The capacity of the Halásztelek well field is about 600,000 m^3 of water per day.

A map of the airfield can be seen in Figure 3. The area consists of a concrete runway 2.5 km long and 60 m wide, and 42 hangars configured in three blocks. Among the hangars are shelters and ammunition magazines. A fuel filling station is located in the southwestern portion of the site. The tank capacity of the station is about 6,000 m^3.

The hydrogeological structure of the area can be seen in Figure 4. About 15-20 m below the surface is the Neocene age clay layer. Above it is a 10-15 m thick Pleistocene-Holocene age sand and sandy gravel layer. The wells of the Halásztelek well field are drilled and screened in this layer. Fine sand and sandy silt, 2-5 m thick, is on the surface.

The groundwater level is about 4-5 m below the surface and it is influenced by the level of the Danube, by precipitation and partly by the Halásztelek well field. The groundwater is flowing to the west and northwest at a rate of 40-50 m/year. Contaminated water is expected to reach the Halásztelek well field in about 12-20 years.

Investigation of the contamination at the airfield was initiated by the Institute for Environmental Management in August, 1991. Based on the results of the investigations, 12

Site Location Map of Tököl Airbase

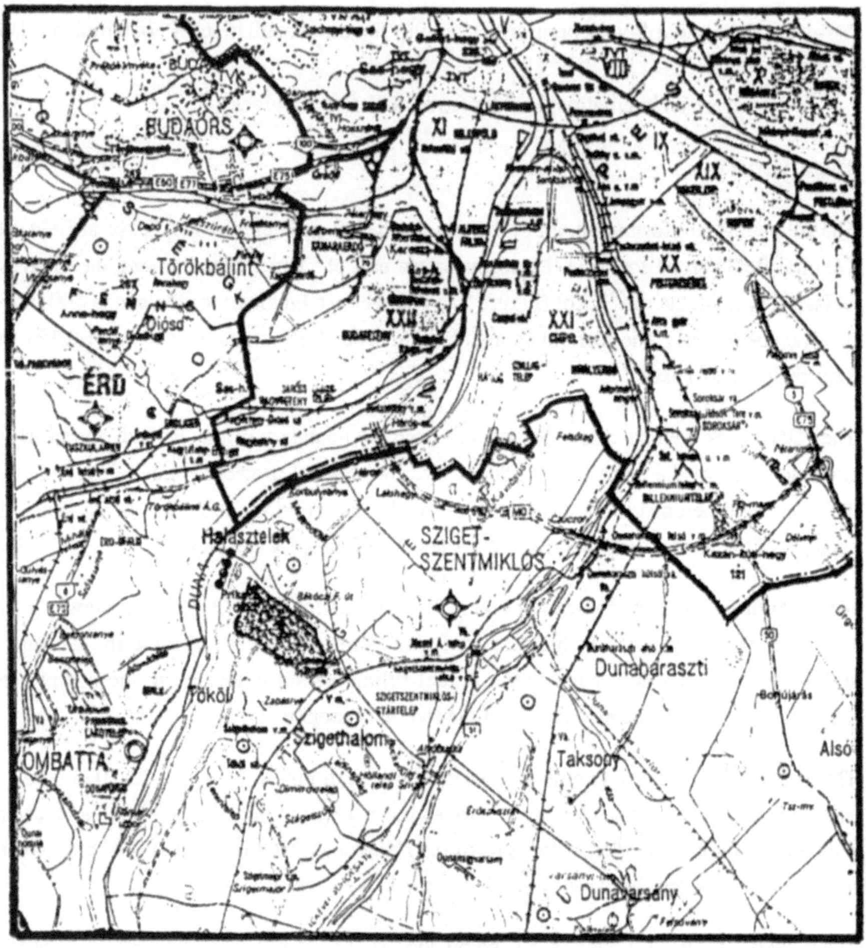

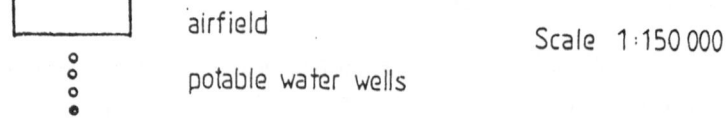

Figure 2

☐ airfield

⋮ potable water wells

Scale 1:150 000

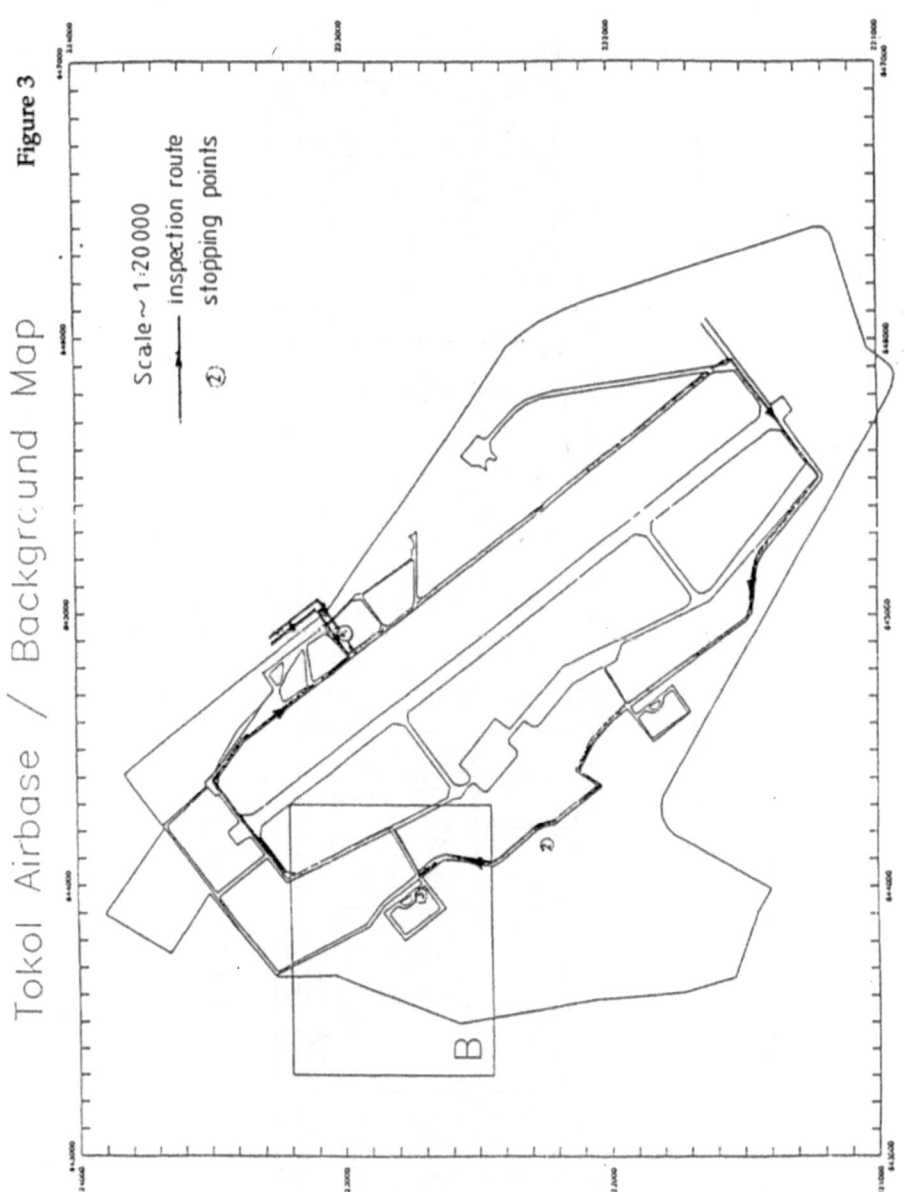

Tokol Airbase / Background Map

Figure 3

Scale ~ 1:20000
inspection route
② stopping points

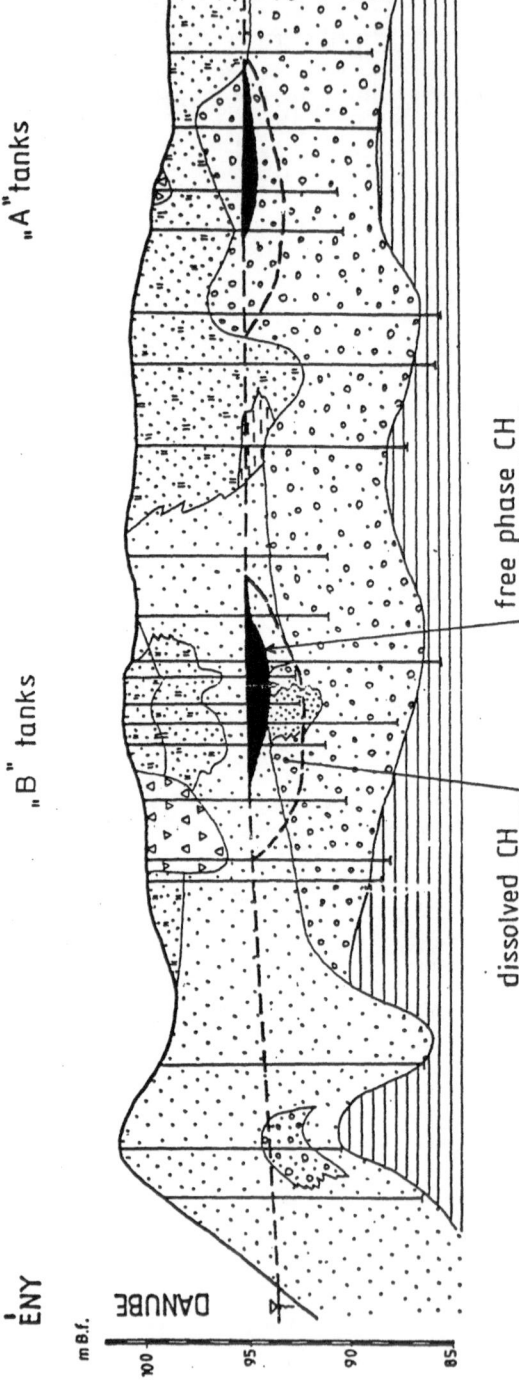

Hydrogeological Sections at Tököl Airport

Figure 4

recovery wells and 46 infiltration wells were drilled in the second half of 1992. The recovery started in 4 places on June 30, 1993.

Recovery of free product and groundwater contaminated with dissolved hydrocarbons was accomplished using the following technique:

- separation of water and free product using 6" small-diameter Filter Scavenger pumps in large diameter recovery wells; coupled with the

- significant depression of the groundwater level with water table depression pumps to increase the flow of free product to the recovery wells.

From November 10, 1990, to June 30, 1993, 279,115 m^3 of contaminated groundwater were pumped out; and 223,853 liters of free phase product were recovered. By April 1994, about 700,000 liters of jet fuel were recovered. Figure 5 illustrates the monthly yields of groundwater and hydrocarbons.

Commissioned by the Danish National Agency of Environmental Protection, a group of Danish firms assisted their Hungarian counterparts in the investigation and remediation of the contamination at the Tököl airfield. This assistance focused on a limited number of tasks (e.g., well construction, pump testing, product recovery, sample collection, chemical analysis, *in-situ* bioremediation) which called for immediate implementation. The assistance was provided in close cooperation with the Hungarian participants.

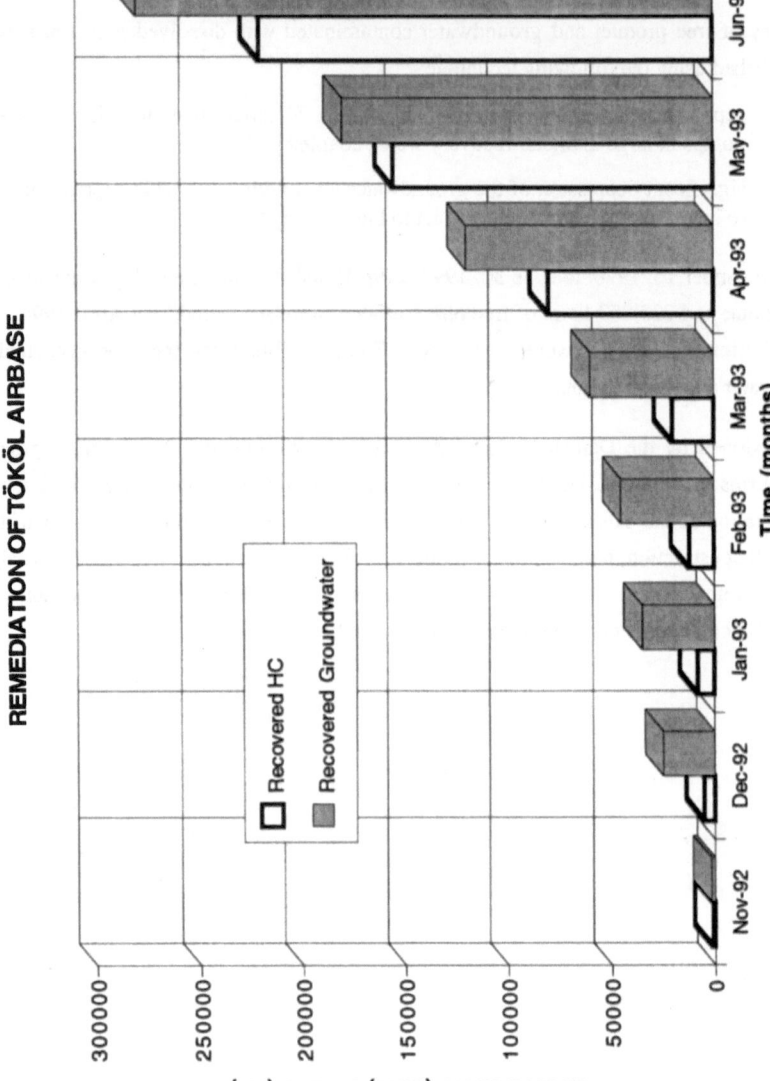

REMEDIATION OF TÖKÖL AIRBASE

Figure 5

ENVIRONMENTAL CONTAMINATION PROBLEMS AT THE

KOMÁROM BASE SITE

Zsolt Sajgó
Institute for Environmental Management
Alkotmány u. 29
H-1054 Budapest
Hungary

The Hungarian government program for site remediation at former Soviet Military bases includes the Komárom Base site. The City of Komárom is located in western Hungary along the Danube River. An army post and a fort built 120 years ago are located at the western border of the town (Figure 1).

The *Árpád Army Post* lies near the center of Komárom, covering an area of about 28 hectares. Between WWI and WWII, family houses and factories were built in this area. Family houses, vehicle repair and maintenance shops, a fuel storage and filling station, and a car wash station were built by the Soviet troops. The present state of the site is one of neglect.

The *Monostori Fort* is located at the western border of Komárom, along the Danube River, covering a total area of about 58 hectares. About 12 hectares of the site are contaminated. There are no residential dwellings nearby. The fort was used primarily for ammunition storage, but a significant number of motor vehicles was also utilized and maintained there. Fuel storage areas, filling stations, vehicle repair and maintenance shops and car wash stations were constructed in support of troop activities.

NATO ASI Series, Partnership Sub-Series, 2. Environment – Vol. 1
Clean-up of Former Soviet Military Installations
Edited by R. C. Herndon et al.
© Springer-Verlag Berlin Heidelberg 1995

Site Location map

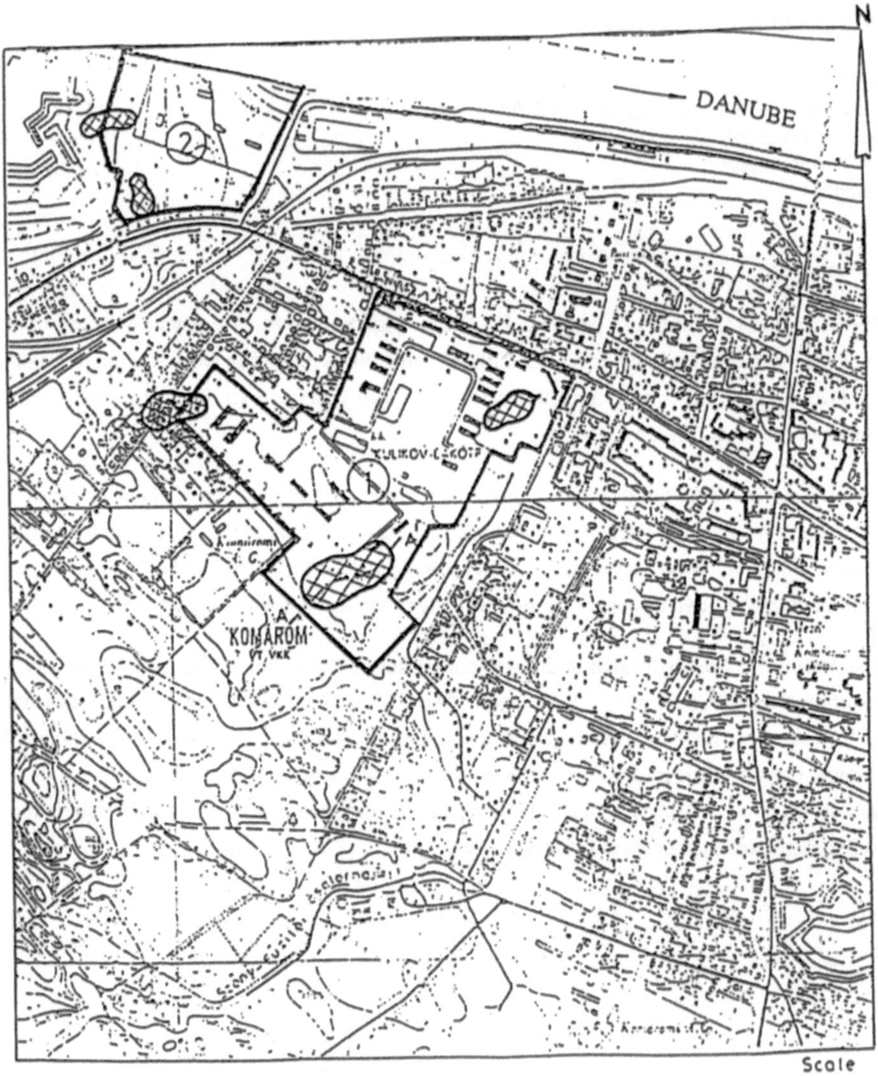

N

DANUBE

Abandoned Former Soviet Military Bases in Komárom, Hungary

Scale
1 : 10.000

1. Árpád Army Post
2. Monostori Fort

Contaminated Area

Figure 1

1. Árpád Army Post

1.1 Survey of Environmental Damage

The site surveys were implemented in two phases. Discussions with the Soviets concerning preliminary investigations took place in the spring of 1991; supplementary investigations were made for planning the remediation activities.

A survey of environmental damages was conducted in 1991. This survey included engineering and geophysical investigations along with chemical analyses of soil, sewage mud and water from the contaminated areas at the Árpád site. Twenty-six sampling locations were established throughout the site. Samples of soil, sewage mud and water were collected primarily in a zone above the groundwater level, approximately 3-4 m below the surface. Samples were analyzed for total hydrocarbon (HC) content using an extraction procedure (CCl4). Many of the samples analyzed were found to contain detectable levels of HC.

In preparation for the remediation activities, bore holes were used as temporary monitoring wells to supplement the results of the preliminary assessment. The basic aim was to determine the state of the contamination, to limit the contamination spatially, and to ascertain the technological parameters needed for the remediation activities.

Information on contaminating sources was collected and sites considered to be critical were identified.

Fifteen bore holes were drilled in 8 areas at the base, ranging between 6.2 and 11.0 meters in depth. All of the bore holes crossed the groundwater storage space of the Pleistocene period and stopped in the Pannonian base. The bore holes were made using the large diameter, dry drilling method. In the overlying and underlying layers, the outer spiral continuous core sampling method was employed (Auger techniques); in loose sand and gravel, a sliming technology was used. Rock samples were collected every 0.5 m above the groundwater level and every 1.0 m below it. Samples of groundwater and hydrocarbons (HC) were taken using wells with diameters of 60-110 mm. Abandoned fuel tanks were also sampled.

Local monitoring well investigations were conducted to determine the hydrogeological parameters. Liquid level measurements of monitoring wells and determinations of hydrocarbon thickness were completed with instruments and skimmers. Investigations were made indicating the contamination of soil and water samples. Chemical tests on soil and water samples (e.g., for total mineral oils, total PAH, benzpyrene, Cr, Ni, Cu, Zn, Cd, Hg, Pb, As, phenols) were made on samples taken every 0.5-1.5 m to the groundwater level at one bore hole at each contaminated area of the site. Additional samples were taken as necessary to evaluate total HC, total aromatic HC and general water chemistry.

1.2 Evaluation of Survey Results

Geological Structure The general geological/hydrogeological structure of the area is shown in Figure 2. Pleistocene sediments consisting of course and fine detritus (e.g., gravel, sand, rockflour, clay) of Danube origin were deposited on a Pannonian surface having a varying ground level. A characteristic feature of the area is a Pannonian block protruding as an upthrust (in the region of *Monostori Fort*).

In the area of the army post, a deposit of detritus about 0.5-6 m is found. Heavy layers (e.g., rockflour, clay) are present in a thickness of 1.5-6 m as well. The course grain layers (e.g., sand, gravely sand, sandy gravel) below the heavy layer have a thickness of 2-6 m. At the base, mostly Pleistocene heavy layers (clay with rockflour, clay) of a maximum thickness of 2.5 m were deposited on the surface of Pannonian clay layers. The Pannonian surface varies between 105-110 m above the Baltic Sea.

Hydrogeological model The Pleistocene gravely sand and sand stores groundwater, and is classified as a medium-to-good aquifer. The Pleistocene clay and Upper Pannonian clayey rocks forming the base of this formation are practically impervious. The overlying layers are slightly aquiferous.

Groundwater is supplied by infiltration and by artificial feeds (e.g., drains). Drainage is toward the local erosion base of the Danube and the channel system of Szöny-Füzitöi.

The groundwater table slopes generally toward the east (NE-E-SE) and is influenced by the Danube River due to its proximity and flow. The groundwater level is at a depth of 3-4 m. The water table was found to be between 107-113 m above the Baltic Sea in June 1992.

1.3 State of Contamination

Visual inspections and preliminary site investigation results show that the following are the primary contaminating sources in the area (Figure 3):

1., 2.	vehicle washing area, fuel filling tank park;
3.	tank park;
4.	boiler house for a bread factory, fuel oil store;
5.	fuel tanks;
6.	boiler house, fuel oil store;
7., 8.	fuel oil tanks.

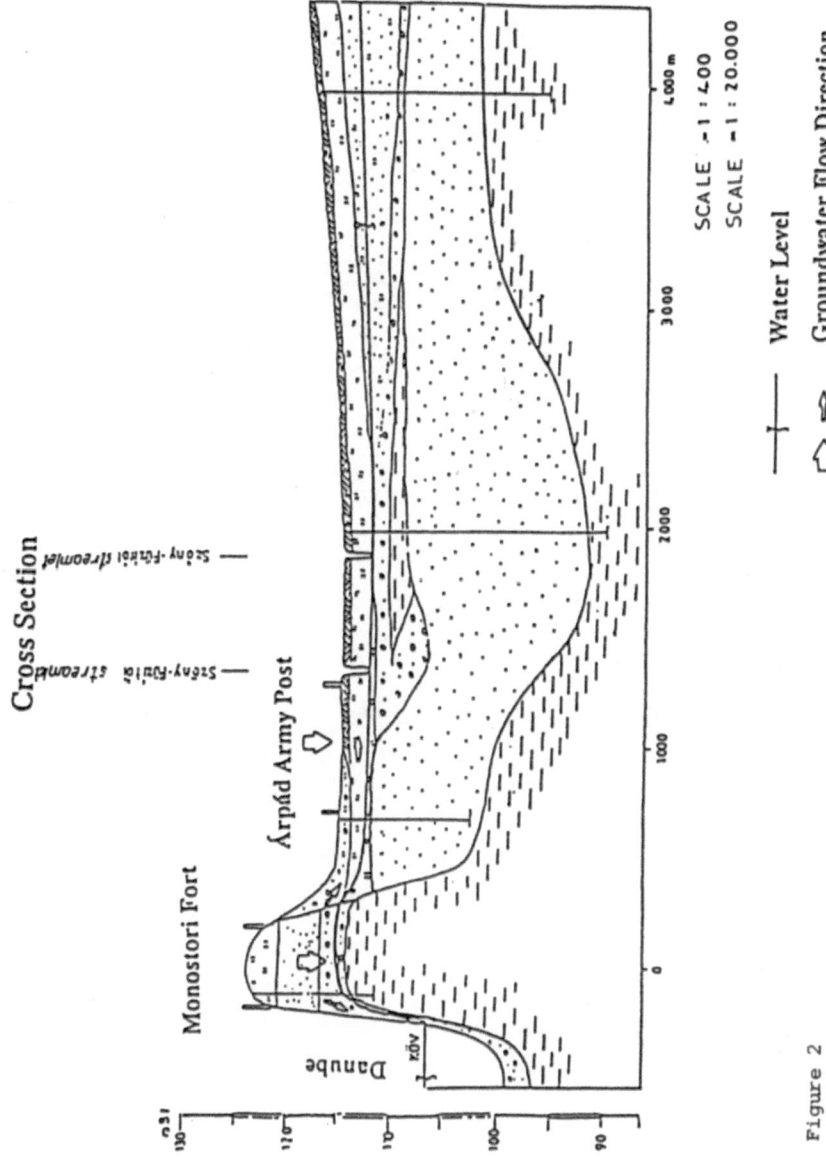

Abandoned Former Soviet Military Bases in Komárom, Hungary

Cross Section

Monostori Fort

Árpád Army Post

Szőny-Füzitői streamlet

Szőny-Füzitői streamlet

Danube

SCALE ~ 1 : 400
SCALE ~ 1 : 20.000

—|— Water Level

⇦ Groundwater Flow Direction

Figure 2

224

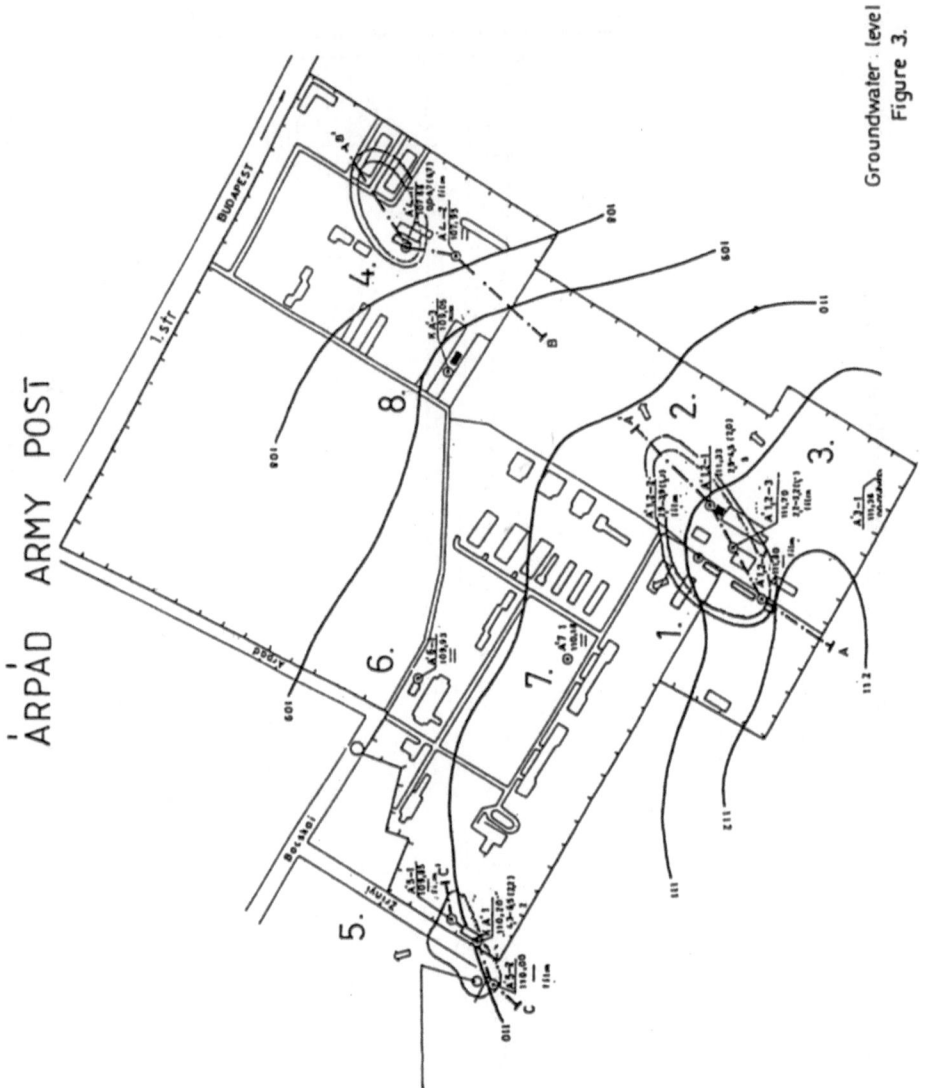

Groundwater level
Figure 3.

The results show that the area was primarily affected by surface HC contamination.

Prior to the site investigation, the volumes of contaminated rock and HC were determined by macroscopic observation, drilling samples (e.g., rock, water, floating HC), and laboratory analysis of water and soil samples. Environmental geological profiles and an areal interpretation were used in this process.

The evaluation revealed that the HC contamination requiring immediate remediation (i.e., reaching the groundwater) is in the vicinity of the car wash station and fuel filling tank park (1, 2), the bread factory boiler house (4), and fuel tanks (5). The HC in area 5 represented a direct danger to the inhabitants since a plume expanded beyond the area of the immediate site.

Soil contamination was observable from 0-5 m in depth with an average thickness of 2 m. One part of the contaminating material infiltrated as far as the groundwater and spread over its surface. At the time of investigation, a thickness of up to 1 cm of HC was observed. The bound HC contamination in the soil was measured between 800-17,000 mg kg^{-1}, with an average value of about 1,500-2,000 mg kg^{-1}. The total dissolved HC content of the groundwater varied between 0.5-8.0 mg L^{-1}.

Other possible contaminants (PAH, toxic heavy metals) were not found in the soil and groundwater samples.

The volume of the residual HC contamination in the soil was estimated using the product of the greatest surface expansion of contamination and the average thickness in this area (see Figure 4). The average thickness was determined from the quotient of the contaminated surface appearing in the characteristic profile and the maximum expansion along its profile. The volume of HC was estimated by taking the average soil concentration into consideration.

The volume of the free HC contamination floating on the groundwater had been estimated by the areal extension of floating HC, the average HC thickness measured in the wells, the density and the free void ratio of the soil into consideration.

It was concluded that primarily gasoline and fuel oil contamination occurred at the site: 82,000 kg in bound form and 354 kg in floating form.

HC contamination has already reached groundwater. Contamination is spreading with the flow of groundwater, degrading the quality of the water in surrounding wells and of smaller surface waster bodies. In area 5 (outside the site) about 100 L of gasoline were pumped from a well.

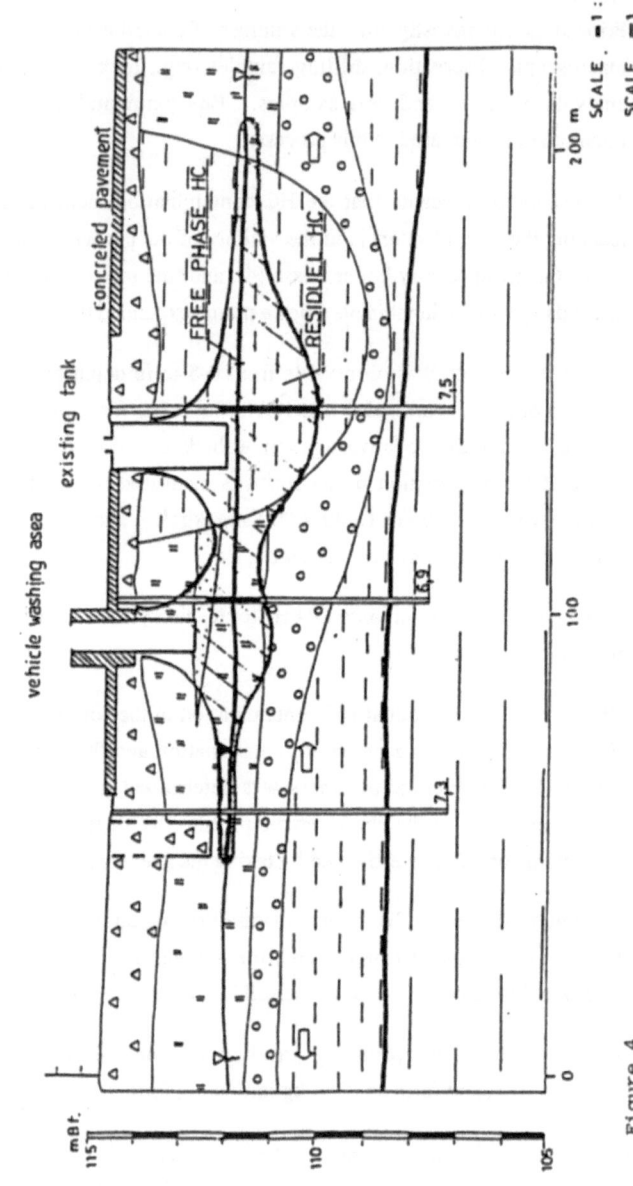

ARPAD ARMY POST
ENVIRONT-GEOLOGICAL CROSS SECTION

Figure 4

2. Monostori Fort

2.1 Survey of Environmental Damage

A preliminary survey of environmental damage at the Monostori site was conducted. This preliminary survey included engineering and geophysical investigations along with chemical analyses of soil, sewage mud and water from the contaminated areas at the Árpád site. Twelve sampling locations were established throughout the site. Samples of soil, sewage mud and water were collected primarily in a zone above the groundwater level, approximately 3-4 m below the surface. Samples were analyzed for total HC content using an extraction procedure (CCl$_4$). Some of the samples analyzed at this site were found to contain detectable levels of HC.

In the next phase, 11 bore holes were drilled to depths of up to 9.6 meters (Figure 5). The drilling and sampling technology was the same as for the Árpád site.

Investigations were made indicating contamination of soil and water samples. Chemical tests on soil and water samples (e.g., total mineral oils, total PAH, benzpyrene, Cr, Ni, Cu, Zn, Cd, Hg, Pb, As, phenols) were taken every 0.5-1.5 m to the groundwater level at one borehole at each contaminated areas of the site. Additional samples were taken as necessary to evaluate total hydrocarbons (HC), total aromatic hydrocarbons, and general water chemistry.

Prior to the remediation activities, bore holes were used as temporary monitoring wells to supplement the results of the preliminary assessment. The basic aim was to determine the state of the contamination, to limit the contamination spatially, and to ascertain the technological parameters needed for the remediation activities.

The drilling of bore holes was initially concerned with the containment of floating hydrocarbon contamination. In contrast with the previous data, floating hydrocarbon contamination appeared also in area 3; investigations were expanded to cover this area as well. Ten bore holes/monitoring wells were drilled in areas 1 and 3.

Local monitoring well investigations were conducted to determine the hydrogeological parameters. Liquid level measurements of monitoring wells and determinations of hydrocarbon thickness were completed with instruments and skimmers.

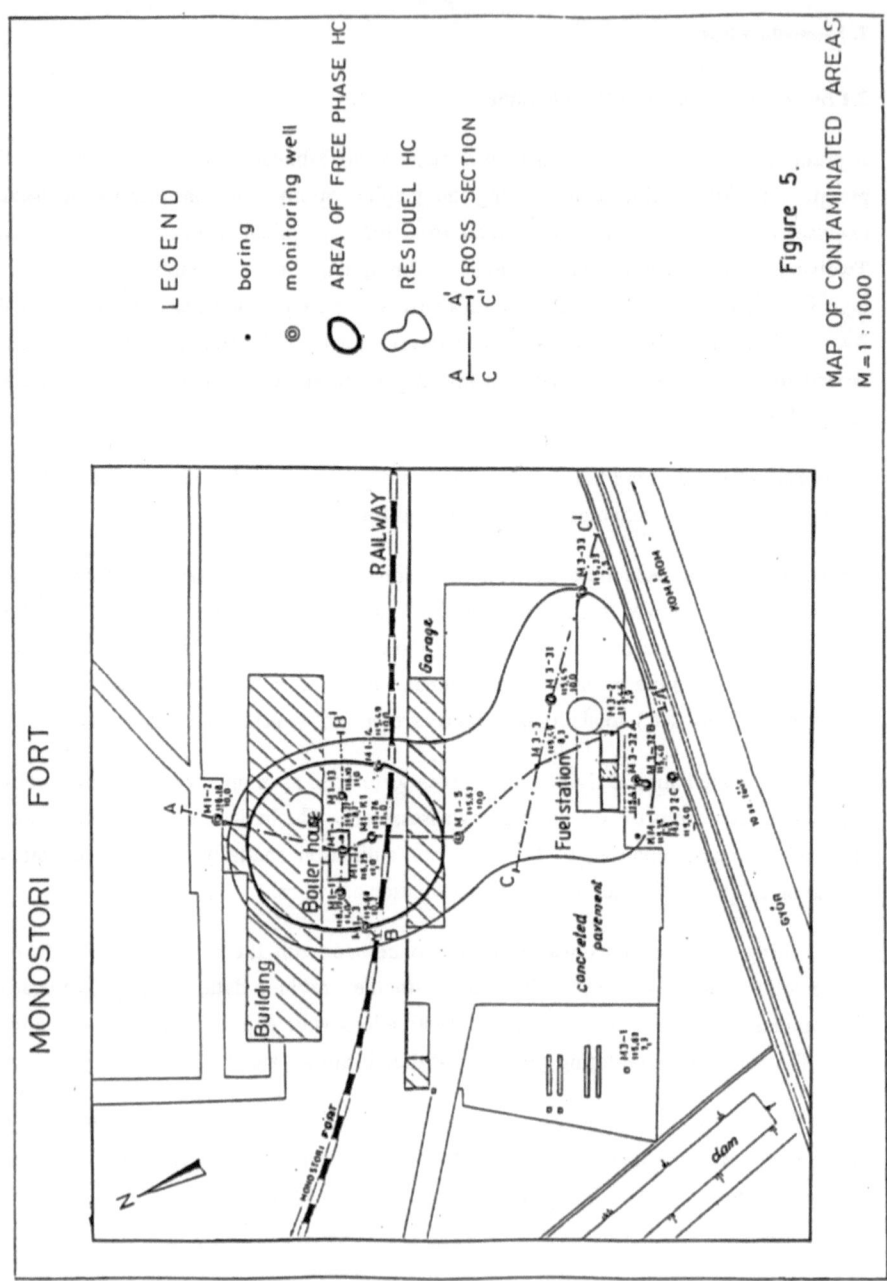

Figure 5.

MAP OF CONTAMINATED AREAS

M = 1 : 1000

2.2 Evaluation of Survey Results

Geologic Structure Pannonian formations have been explored which were about 0.5-4.0 m thick at a depth of 6.0-10.0 m below the surface in the vicinity of the fort. The material is mostly clay, and less frequently clay with rockflour and sand. The surface is uneven, varying between 106-109 m above the Baltic Sea, elevating toward the Monostori Fort (NW) and showing a regional sloping towards the Danube (NE). In the middle of the investigated area, a slight, local, dish-like depression is present.

Above the Pannonian surface, the decisively fine detritus formations from the Pleistocene age are found over the entire area. These vary between 0.0-5.3 m in thickness; the compositions of these formations are rockflour, sand with rockflour, and sand. On the Pannonian land surface, there occasionally occurs sand with gravel spread and sandy gravel 0.7-1.0 m thick.

Hydrogeological model In the area of the investigation, Pannonian formations are almost impervious.

The detritus series from the Pleistocene age stores groundwater. The fine grained lower layers have a weak, eventually medium aquiferous feature. The parts with coarse detritus appearing as spots in some places have a medium-to-good aquiferous features. The series with fine and course detritus appearing in the upper parts also have a medium-to-good aquiferous features. Based on grain distribution, the following coefficients of permeability are characteristic: rockflour, sand with rockflour 3.5×10^{-7} - 9.6×10^{-6} m s^{-1}; and gravely sand, sandy gravel 1.6×10^{-5} - 1.4×10^{-4} m s^{-1}.

Based on *in situ* pumping-backfill investigations, the coefficient of permeability of rockflour and sand with rockflour soils is between 3.4-6.3×10^{-6} m s^{-1}. It should be noted that in this same soil the coefficient of permeability for hydrocarbons (well M1-1) was 4.6×10^{-7} m s^{-1}.

Groundwater at the site and in the area in general is supplied by infiltration of rainwater in the vicinity of the Monostori Fort. The drainage occurs, to some extent, toward the southward oriented alluviums, but more so toward the erosion base of the Danube River.

The position of the groundwater reflects the fact that it slopes from the Monostori Fort area toward the drains. According to reports on August 25, 1993, the water table has stabilized between 110.18-109.76 m above the Baltic Sea at a depth of 5.27-6.44 m below the surface.

2.3 State of Contamination

In the investigated area, two concentrated contamination sources can be identified:
1. Fuel oil tank in the boiler house; and
2. Gas oil tank of the filling station.

In addition to this contamination, these areas were also affected by an areal surface of HC contamination.

According to data from previous investigations (1991, 1992), an estimated 13,000 L of floating HC contamination was on the surface of groundwater in the vicinity of the first contaminating source. Accompanying this and expanding to Area 3 are 16,680 m^3 of HC contaminated soil, or about 30,024 kg of HC material. Based on the evaluation, it became evident that the contamination requiring remediation was located in the area of the filling station no. 3 beyond the boiler house area (Figure 6).

The volume of HC contaminated soil was 20,100 m^3; 138 m^3 in bound form (soils), and the volume of floating HC is estimated to be 12 m^3. The thickness of the floating HC varied between 20-1300 mm, and the bound HC content varied between 800-38,000 mg kg^{-1}. The dissolved HC content of the groundwater was between 1.0-28.0 mg L^{-1}.

Based on the investigation results, it can be stated that the area was affected by HC contamination of great quantity. Part of this is surface contamination, but the overwhelming majority is subsurface contamination influenced by groundwater flow either bound in the soil, dissolved in groundwater, or floating on the groundwater.

Areal propagation of dissolved hydrocarbons can be estimated using the spreading area of bound hydrocarbons in the deeper soil zone. The contaminated groundwater propagates at a rate of 8-15 m yr^{-1}.

Floating contamination must be removed in order to prevent HC contamination from endangering the water quality of the Danube. This contamination also moves in the direction of the Danube on the surface of groundwater at a slower rate than the dissolved contamination. Of course this speed is related to the slope of the immediate area which increases close to the Danube, where the rate of spread of contamination may approach 30 m yr^{-1}.

The site is located in a developed region. Future plans for the site include a tourist area and a historic monument. This requires the immediate implementation of remedial activities to a suitable level.

A–A' ENVIRONMENT–GEOLOGICAL CROSS SECTION

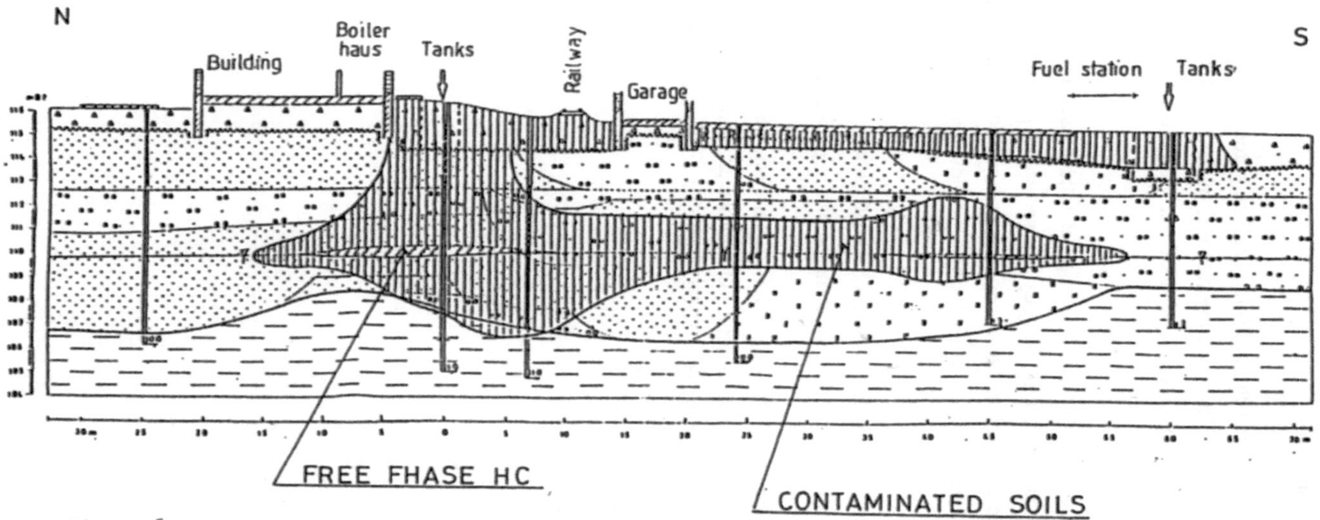

N

S

Building

Boiler haus

Tanks

Railway

Garage

Fuel station   Tanks

CONTAMINATED SOILS

FREE FHASE HC

Figure 6

3. Concept of Remediation

Groundwater supplies within the inhabited area are potentially endangered. The immediate goal of the remediation efforts was the removal of floating hydrocarbons. Although this reduces the risk of direct damage, it does not exclude the necessity of complete site remediation.

Total remediation of environmental damages is time-consuming and costly and, therefore, may be prohibitive at this time in Hungary. However, the partial elimination of hydrocarbon contamination, the removal of the free-phase hydrocarbon contamination floating on the surface of groundwater, and the purification of groundwater contaminated with hydrocarbons can be realistically achieved.

Goals were set which stated that no free-phase HC should remain and that the dissolved HC concentration should not exceed a concentration of 2 mg L^{-1} after purification.

Initially, the elimination of floating HC contamination was achieved, eliminating the main source of dissolved HC contamination. Groundwater found in the area of the floating HC should be purified with phase separation, filtration and ventilation prior to either surface release or backfilling into the soil.

ENVIRONMENTAL PROBLEMS ASSOCIATED WITH FORMER

SOVIET MILITARY INSTALLATIONS IN POLAND

Zbigniew Kamienski
Control Department
State Inspectorate for Environmental Protection
Wawelska 52/54
00-922 Warsaw
Poland

Introduction

The last Russian soldier departed Poland on September 17, 1993. The overall presence of the Russian Army stationed in this country since World War II totaled about 60,000 soldiers deployed in 59 sites covering jointly about 70,000 hectares of land. The sites included air fields, fuel depots, ammunition storage facilities, testing grounds and a naval port.

The entire set of issues pertaining to the ecology of the areas formerly occupied by the Army of the Russian Federation (ARF) has been committed to the State Inspectorate for Environmental Protection by the Minister of Environmental Protection, Natural Resources and Forestry.

Until 1990, the area occupied by the ARF could not be inspected except on rare occasions due to the, then binding, regulation prohibiting inspections teams from entering the sites. At that time, inspections were only performed either in the case of a complaint brought by the local population or under conditions of extraordinary threat to the environment, with permission granted by the Headquarters of the Northern Group of Soviet Forces (NGSF). In 1990, the State Inspectorate for Environmental Protection (SIEP) performed environmental inspections of nine selected ARF sites including large fuel depots, air fields and testing grounds. As a result, it has been determined that the primary threats to the environment involve the pollution of land and groundwater by petroleum derivatives. The ARF illegally abused the environment, violating both the standards and environmental requirements. Additionally, neither fees for using nor penalties for polluting the environment were paid by the ARF. It was considered necessary to perform detailed environmental inspections at all sites, to subject

NATO ASI Series, Partnership Sub-Series, 2. Environment – Vol. 1
Clean-up of Former Soviet Military Installations
Edited by R. C. Herndon et al.
© Springer-Verlag Berlin Heidelberg 1995

further ARF stationing and withdrawal operations to permanent supervision by environmental protection authorities, and to assess the overall ecological damage resulting from the stationing of the Russian Army in Poland.

In the beginning of 1991, the SIEP initiated a study designed to identify and assess the ecological damage brought about by the Army. In the second half of 1991, following the presentation of results obtained at some of the examined sites, the NGSF Headquarters first impeded activities and later entirely prohibited the SIEP inspectors from entering the then occupied sites. Due to numerous actions aimed at resolving the problems directed through the Plenipotentiary of the Polish Government for the Matters Concerning Stationing of the Russian Army in Poland as well as political actions taken through the intermediary of the Ministry of Foreign Affairs, the SIEP persevered with the above study. The SIEP also examined areas beyond the limits of respective sites since adverse effects of land and groundwater pollution had been detected beyond the area actually used by the ARF.

After the signing of the Russian Army withdrawal agreements in Moscow on May 22, 1992, joint Polish-Russian inspections were made at all sites, resulting in the establishment of protocols concerning the environmental status of these areas. These protocols constitute integral parts of the delivery-acceptance protocols for these sites. The SIEP has also exercised current ecological supervision of the withdrawal operation of the Russian Army. Recommendations were then directed to the NGSF commander-in-chief, emphasizing the necessity to prohibit the following activities:

- the burial of chemicals, waste and other materials;
- the transportation of chemicals and waste without prior clearing with the Voivodeship Inspector for Environmental Protection;
- the discharge of petroleum derivatives and other liquid chemicals to the ground and waters;
- attempts to camouflage existing ecological damages; and
- the dismantling of facilities and installations which might result in environmental contamination.

Supervision of the Russian Army withdrawal by the SIEP was impeded by Russians to the point of issuing a temporary ban on the inspection of some of the sites. On some occasions, the SIEP inspectors were not allowed to examine certain portions of the sites. However, persistent action undertaken jointly by other bodies of governmental and local administrations enabled, among other things, the detection of illegal disposal of chemicals and waste materials.

The SIEP representatives took part in the negotiation and elaboration of the Polish-Russian agreement on the withdrawal of the ARF from Poland. Consequently, the issues of environmental protection have been distinctly addressed in the documents signed, though initially there was a clear attempt on the part of the Russian delegation to regard environmental matters as unimportant. The assessment of ecological damage performed at some of the sites turned out to be of primary importance during the negotiations concerning financial and ownership matters.

In November of 1992, a Team was appointed by the Minister of Environmental Protection, Natural Resources and Forestry to coordinate the investigation and restoration of areas formerly occupied by the ARF. The Team was headed by the Chief Inspector for Environmental Protection. The Team includes, among others, Deputy Voivodes from Voivodeships where major Russian forces were stationed. The role of the Team is to coordinate the entire range of efforts involving establishment of a time-table for environmental safeguarding and remediation operations, optimization of financial expenditures to this end, selection of appropriate environmentally-friendly restoration techniques, and comprehensive technological and organizational assistance.

Scope of Investigations Conducted to Date

The environmental inspections performed to date by the SIEP have provided evidence of the existing and/or potential threat of the diverse pollution and contamination hazards to the environment at 35 sites. Taking into account the elevated costs of investigation necessary to identify and assess the damages, 21 sites were qualified for further analysis, including those where pollution and damage were expected to be most severe. Listed among those sites were all of the air fields with large fuel depots, all testing grounds, ammunition storage facilities and selected sets of barracks. The results of investigations have confirmed the validity of the adopted scope of studies.

The following aspects were studied at the sites:

- pollution of soil and groundwater by petroleum derivatives and other chemicals;
- pollution of surface water;
- damage to and pollution of land surface;
- contamination by poisonous combat chemicals;
- radioactive contamination; and
- damage to forests.

The data collected at the sites were derived from the following investigations:

- Aerial photographs of the sites at the scale 1:10,000. The photographs provided the basis for a preliminary field survey since existing maps were outdated;

- Field surveys;

- Hydrological studies including drillings of exploratory boreholes and sight-holes (piezometers). The latter were used for the purpose of sampling soil and water to be further physico-chemically analyzed and for determining the direction and velocity of the groundwater flow;

- Physico-chemical and analytical study; and

- Dosimetrical studies.

Within the framework of studies designed to identify and assess ecological damage, a set of urgently needed actions has been defined in order to halt situations that endanger the environment at respective sites. The actions will be primarily directed to eliminate pollution sources such as: petroleum derivatives floating on the groundwater surface or at the top of impervious layers; refuse dumps; sites contaminated by harmful and toxic chemicals; and waste discharged to surface waters. All of these sources threaten the environment with further pollution or contamination of water supplies. Based upon the results of the study and analysis of local conditions, a list of threats has been elaborated as well as a ranking of urgent actions aimed at environmental safeguarding and remediation.

The studies made at these sites provided a sufficient basis for decisions concerning the necessity of immediate action aimed at safeguarding and restoring the environment and at target-oriented remediation undertakings. Additional investigation may eventually be required to ensure the implementation of an effective, long-term safeguarding and remediation plan.

Overview of the Status of Environmental Damage

Studies performed indicate that it is the pollution of soils and groundwater that pose the greatest environmental threat. The main pollution sources are petroleum derivatives containing various hydrocarbons and chemical mixtures.

These compounds pollute the ground, producing on multiple occasions a distinct layer (up to several meters thick) of fuel floating on the surface of groundwaters or at the top of impervious layers. Petroleum-derivative contamination was found at all the examined sites, while the size of the polluted areas and content of the petroleum derivatives varied greatly among the sites.

Other equally hazardous types of pollution resulted from illegal waste dumping. Soils adjacent to dumping sites were found to be contaminated with harmful substances. Such waste dumps are situated within 19 sites.

Environmental threats originate also from the harmful and toxic chemical pollution found at five of the sites. The pollution was caused by storage of chemicals, field-testing of chemical warfare agents, and burning of unknown substances.

The contaminated grounds comprise approximately 1.0% of the examined land. These grounds should be excluded from future agricultural use. Whether they can be used for other purposes depends on the success of remediation work. On about 10% of the investigated area, pollution has adversely affected the status of groundwater, which can not be regarded as suitable for drinking or other domestic uses. This large amount of polluted groundwater results from pollution dispersion through groundwater flow. This pollution poses a major threat to water supplies and to surface water.

The following types of pollution and environmental damage can be treated as relatively less hazardous to the environment:

- pollution of surface water reservoirs and of waterways (due to their small number);

- land disturbances encountered within working fields of testing grounds (physical degradation of the top soil has been found to exceed 25% of the investigated land, but no chemical pollution has been found there); and

- damages of various degrees to the forested areas (the forested areas represent about 63% of the land in question).

Neither contamination by poisonous chemical weapons nor radioactive contamination was found in the course of the investigations made to date.

The scale of existing ecological damage is best illustrated quantitatively. Within the ARF stationing sites subject to investigations, the following have been polluted and/or damaged:

- soils by petroleum derivatives, in the amount of about 18.4 million m^3, within an area of about 406 hectares; the most adverse effects have been found in the saturated zone having a volume of about 5.17 million m^3;

- a groundwater aquifer in an area of about 6,500 hectares, having a total volume of about 145 million m^3; the water is not suitable for drinking or for other domestic uses;

- waste dumping sites covering about 98 hectares;

- soils contaminated with toxic chemicals in the amount of about 0.97 million m^3 in an area of about 22 hectares;

- surface water reservoirs of about 17.5 hectares;
- soil-vegetation layer within an area of about 15,330 hectares; and
- forested areas of about 38,100 hectares.

Cost Estimate of Environmental Damage

A cost estimation of the environmental damage was an essential element of the investigation performed and, to this end, an especially elaborate method was used. Overall ecological damage to the investigated sites has been assessed to be about $2.4 billion (USD).

This assessment of ecological damage constitutes an equivalent of costs that would be incurred in order to completely restore the area, that is to the point of attaining its original status. However, we have to assume that there will be neither the possibility nor the necessity for achieving that status. The successful elimination of the damages will, to a great extent, depend upon the management method applied. The approach will vary depending on the intended future use of the land, whether it be for recreation, agricultural use, settlement, or industrial usage.

For the immediate implementation of urgent operations aimed at safeguarding and the remediation of the sites, a total of about $104 million (USD) is needed.

Threats and Urgent Safeguarding and Remediation Operations

Field operations involved in cleaning the environment can be divided into safeguarding and remediation operations.

The environmental safeguarding operations which consist of halting pollution migration need to be done irrespective of land management type. These safeguarding operations can be divided into two categories. The first embraces operations intended simply to prevent the further spread of pollution, without actually diminishing the degree of environmental pollution. The other embraces operations designed to remove or treat polluting substances, such as removal of petroleum derivatives from the ground in order to reduce the environmental pollution. The latter operations constitute the dominant type of safeguarding work, as well as the first phase of remediation. Both groups of operations mentioned above have been labeled "urgent safeguarding and remediation operations."

Further environmental remediation will be strictly related to the intended land use of the site, since it is the land use objective that will define the scope of remediation work. Procedures in specially protected areas will be different from those designated for industrial use. It may even happen that further remediation work at certain sites will be entirely unnecessary.

The investigation of sites has shown that the most urgent safeguarding operations embrace the following activities:

- Removal of petroleum derivatives from a floating layer in the ground;
- Waste dump remediation;
- Remediation of land contaminated by harmful and toxic chemicals;
- Removal of waste discharged into surface water bodies. There is a well-grounded suspicion that some metal containers holding chemical substances may corrode and contribute to surface water pollution.

Remediation Operations

In accordance with the adopted assumptions, the type of remediation work to be performed at individual sites will largely depend upon the targeted land use. The targeted land use will also define the extent of environmental treatment. It is envisioned that the polluted area will be classified into one of three groups differing in the level of admissible pollution.

The first group embraces legally protected terrain such as national parks and reserves, spa water supply zones, areas that feed utilized aquifers, and protection zones of groundwater supply points.

The second group encompasses agricultural lands (including fields of grain, pastures, orchards), forested areas, residential areas, recreation areas and general amenity lands.

The third group includes air fields, testing grounds, fuel depots, industrial enterprises, roads and areas in support of industrial plants.

The type of soil is an essential consideration influencing the scope of remediation work. The treatment of permeable soils such as sands of various grain structure, gravel or sand-gravel mix is much easier than that of compact soils such as clays and silts. It often happens that due to elevated costs and sophisticated technologies required, the further treatment of compact soil is totally abandoned.

Advancement of Safeguarding and Remediation Operations

Parallel to the assessment of ecological damage, the SIEP has developed the following guidance materials within the framework of preparation to eliminate the damage:

- "Ranking list of urgent safeguarding and remediation operations required at the former ARF stationing sites";

- "Principles of bidding for performing the safeguarding and remediation work in areas where ecological damage was found due to the ARF stationing", and "A catalogue of firms offering services in the field of safeguarding and remediation in areas where ecological damage was found due to the ARF stationing"; and

- "Guidelines concerning the degree of remediation of soils and groundwaters abandoned by the Russian Army and polluted by petroleum derivatives and other chemicals substances".

These guidance materials are necessary for determining the order of operations for technically and economically optimizing the implementation of environmental safeguarding and remediation work.

At this time, all of the preparatory work needed to enable implementation of the above operations has been completed. The rate of advancement of the operations will depend upon available financial resources.

A relatively small amount of environmental safeguarding work has already been done. In the current year, 16 projects will be initiated in conformity with the ranking list of urgent operations for safeguarding and remediation of the environment.

ENVIRONMENTAL ASPECTS OF REUSING FORMER SOVIET ARMY BASES IN SLOVAKIA

Elena Fatulová
Ministry for Environment of the Slovak Republic
Water Protection Department
Hlbok 2
812 35 Bratislava
Slovak Republic

Daniel Geisbacher
Slovak Inspection for Environment
Karloveská 2
842 22 Bratislava
Slovak Republic

Within the framework of the agreements made between the former Czechoslovakia and the former Soviet Union concerning the mutual compensations associated with the occupation of the Soviet Army (SA), arrangements were made to assess the environmental damages at the many former military installations. A Czecho-Slovak commission was formed which would work under the direction of and in cooperation with the former Federal Ministry of Defense.

From July through September of 1990, visual inspections were performed in Slovakia at 87 sites in 18 different installations where SA units had been stationed. These inspections were aimed at the identification and characterization of the type and extent of environmental damage at the sites in order to identify means of remediation and the eventual re-use of these facilities. Emphasis was placed on contamination of soils and groundwater. Inspections were performed on sites with visible evidence of surface contamination and/or subsurface pollution, mainly of petroleum products, at waste sites, fuel deposits and bunkers, and at other areas at these sites.

The 18 inspection projects (essentially the same for each site) were completed in 1991. These nationally coordinated efforts served to identify sites which were in need of emergency remediation measures and to estimate cleanup costs.

NATO ASI Series, Partnership Sub-Series, 2. Environment – Vol. 1
Clean-up of Former Soviet Military Installations
Edited by R. C. Herndon et al.
© Springer-Verlag Berlin Heidelberg 1995

Soils and groundwater were sampled and analyzed in order to determine the composition of the waste deposits. Soil samples were taken at 1-meter depth intervals from the surface to the groundwater level; these data were used to determine the depth horizons of the contaminants within the soil. The soils were found to contain oil substances (collectively measured as NES - non-polar extractable substances), heavy metals, and specific organic substances.

Water samples were taken from hydrogeological bore holes at the centers of pollution, as well as from wells and springs in the adjacent areas. Two to three areal samples were taken at each site at various times.

Samples were analyzed for various trace elements including phenols, aromatic and chlorinated hydrocarbons, PCBs, and bacterial content. In environmentally sensitive areas, analysis of landscape relief, karst phenomena and flora were also analyzed for changes induced by the SA.

Prior to the evaluation work, acceptable pollution limits for soil and groundwater were determined by the state administrative agency. In defining these limits, water management importance, as well as background levels of contaminants in surrounding non-contaminated areas, were taken into account. CSS (Czechoslovak Standards) were utilized for drinking water concentrations of conventionally accepted indicators. Soil contamination was measured using the NES content; limits varied from 100-600 mg kg^{-1}. These standards were used to designate those sites needing remediation, and those which could be considered "clean".

Final reports for each site contain data on concentrations and composition of contaminants in soils and groundwater. Data are presented graphically within these reports, and proposals for remediation have been made. Results of these investigations are presented in Table 1.

As indicated, 14 out of the 18 installations will require remediation efforts. For the most part, the contamination was composed of petroleum products, caused by leaky fuel bunkers and piping, as well as by unprofessional and irresponsible handling of these materials.

The total volume of the contaminated soils for all sites is estimated to be 250,000 cubic meters, while contaminated groundwater was found over approximately 20 square kilometers.

Cost estimates for remediation reached 931 million Slovak crowns (Sk) (according to average costs at the 1993 price level). Of this amount, more than half (621 million Sk) went to remediation of the Sliac-Vlkanová installation. Other major remediation sites include Nemsová, Lest', Komárno and Ruzomberok, with costs ranging from 30-90 million Sk per installation. Additional sites, evaluated as having less pollution than the aforementioned sites, required an average of 15 million Sk per site.

Table 1

#	Location	Restoration needs	Cost of restoration (mil Sk)		Amt. invested (mil Sk)
			Est.	Project	
1	Sliac-Vlkanova	-decontamination of soils, (170,000 m^3) (limit 200 mg kg^{-1} for Vlkanová, 500 mg kg^{-1} for Sliac) -construction of drainage system and treatment stations for sustainable protection of groundwater -decrease NES content, aromatic chlorinated hydrocarbons, PCB in groundwater to the level of category B of Recommendation -liquidation of pollution sources, building wastes	621.00	640.00	152.58
2	Nemsová	-liquidation of objects on the area of 32,000 m^2 -decontamination of polluted soils by NES (750 m^3) -decontamination of the concrete areas and panels (6400 m^3) contaminated by NES (limit for extraction 200 mg kg^{-1}, for cleaning 500 mg kg^{-1}) -restoration of the ground waters (decrease oil substances to 0.5 mg L^{-1} NES)	90.00	37.00	31.26
3	Lest'	-removal of fuel tanks, cleaning of objects -soil decontamination (11,700 m^3), decrease NES to 400 mg kg^{-1} -restoration of groundwater (NES, PCB, aromatic and chlorinated hydrocarbons - limit to category B of Recommendation	34.20	81.5	46.95
4	Zvolen	-restoration of soils (3,500 m^3), decreasing NES to 500 mg kg^{-1}	4.52	-	-
5	Komárno	-soil restoration (33,300 m^3), limit 600 mg kg^{-1}NES -restoration of groundwater, limit to category B of recommendations	56.85	-	-
6	Ruzomberok	-liquidation of structures, recultivation of 18 hectares land -soil restoration (17,300 m^3), limit 500 mg kg^{-1}NES -restoration of groundwater, limit 0.05 mg L^{-1}NES	31.02		
7	Nové Zámky	-soil restoration volume of 1,900 m^3 - limit 200 mg kg^{-1} NES -restoration of groundwaters - limit 0.05 mg L^{-1} NES	15.56	-	-

Table 1 (continued)

8	Nové Mesto nad Váhom	-object restoration (bridge for repairs) -soil restoration (7,300 m^3) - limit 600 mg kg^{-1} NES -restoration of groundwater	8.48	-	-
9	Rimavská Sobota	-soil restoration (11,000 m^3) to limit 600 mg kg^{-1} NES -restoration of groundwater	14.96	-	-
10	Roznava	-restoration of pollution sources -soil restoration (5,375 m^3) - limit of 600 mg kg^{-1} NES	4.63	-	-
11	Jelsava	-restoration of pollution sources -soil restoration (380 m^3)	0.98	-	-
12	Stúrovo	-soil restoration (3,550 m^3) - limit 200 mg kg^{-1} NES -soil restoration(5,300 m^3) - limit 300 mg kg^{-1} NES	6.86	-	4.46
13	Voderady	-liquidation of pollution sources -soil restoration (144 m^3- limit 200 mg kg^{-1} NES	0.18	-	0.17
14	Castkovce	-soil restoration (230 m^3) -restoration of ground waters	1.63	-	1.63
15	Vrútky	-no areal contamination proved	-	-	-
16	Kezmarok	-no areal contamination proved	-	-	-
17	Michalovce	-no areal contamination proved	-	-	-
18	Skalka nad Váhom	-no areal contamination proved	-	-	-
	All locations	-monitoring			
	Sum		931.56		237.05

Sk = Slovak crowns

Remediation began at the end of 1992; efforts were concentrated on those areas with a high degree of contamination, those posing a serious threat to the environment, and those with significant risk to water resources.

The first goal of the remediation efforts was to remove the existing sources of contamination (i.e., tanks, piping, containers, pumping equipment, storage facilities, dumps, etc.) Subsequent treatment of the soils and groundwater was dependent upon the intended future land use and water management importance. Economic and technical factors were also taken into account. Technical aspects of the cleanup were essentially the same for all sites.

Soil cleanup consisted of removal of soils to concrete-lined treatment sites, where soils were treated through bioremediation techniques. Contaminated groundwater was pumped from hydrogeological bore holes constructed at the center of the polluted sites, then purified and discharged into surface waters. Purification processes were overseen by government authorities.

The greatest remediation efforts were carried out at Sliac-Vlkanová, Nemsová and Lest'; the following sections describe the efforts at these sites.

Sliac-Vlkanová

This site is significant because it is located within the second protection zone of the thermal mineral waters used in the spas of Sliac and Kovácová. The site is located on the right bank of the Hron river; contaminated water is continually draining into the river. The surrounding geology is a complex mixture of Quaternary and Neocene clay, sand and gravel. Thickness of this layer varies from 5 meters in Vlkanová to 130 meters in Sliac. The bedrock is neo-volcanic rock over Mesozoic carbonates, which are mineral water collectors.

Until 1990, this site was used by the former SA as a military airfield; support facilities included fuel storage vessels, garages, barracks, watch towers, etc. Army activities resulted in extensive pollution of soils, rocks, and groundwater, primarily with hydrocarbons and halogenated derivatives. Pollution of soils is patchy, located at 14 different points. The total polluted area at Sliac-Vlkanová represents 130,000 m^2, or about 170,000 m^3 of soil. Contamination is found at various depths, even to the groundwater level. NES content varies significantly; maximum NES found was 16,000 mg kg^{-1}.

Groundwater pollution covered an area of approximately 5 square kilometers. NES ranged from 0.05 mg L^{-1} in marginal areas to tens of mg L^{-1} at the centers of pollution. At these centers, contaminants were found to have a depth of 0.2-1.3 meters. Groundwater contaminants included organic pollutants, hydrocarbons and halogenated derivatives, aromatic hydrocarbons, and PCBs.

Pollution in this area has been known since the early 1980s. At that time there was an oil layer 1-4 meters in depth at the groundwater level. From 1981-1991, 336,278 liters of petroleum substances, primarily kerosene, were pumped from the groundwater surface. Yet the total effect of the purification was not significant until after the departure of the SA and the subsequent removal of pollution sources, such as fuel tanks.

In October of 1992, a final remediation project proposal was outlined which would complete the cleanup work by 1997. However, lack of funding has caused significant delays at this site, and completion in this time frame is very unlikely.

Strict guidelines have been set for the extraction and purification of soil. Allowable NES levels have been set at 200 mg kg^{-1} for Vlkanová and 500 mg kg^{-1} for Sliac. For groundwater, limits have been adopted which are equivalent to the "level B" of "The Netherlands Standard". The Ministry of Environment of the Slovak Republic accepted this as a basis for "Recommendations of the Slovak Commission for Environment for application of indicators and directions for restoration of polluted soils and groundwaters".

After the removal of debris, scrap metal, and various containers, (much of which was used as fill material), on-site remediation methods of soil decontamination were employed. Contaminated soils were excavated and relocated to concrete-lined holding tanks, where they were sprayed with a water-based mixture inoculated with bacteria obtained from the microbial bank at Masaryk University in Brno. After homogenization, the soils were formed into pyramid-shaped mounds about 2-2.5 meters high. The mounds were aerated every 4-6 weeks, and irrigated with a water/bacteria/nutrient mixture according to the rate of degradation of the oil substances, which was monitored continually. Once contaminant concentrations dropped below the threshold levels, the soil was used to back fill the excavation sites. About 13,000 cubic meters of soil have been treated.

Groundwater pumped from the 25-30 hydrogeologic bore holes passes through a treatment facility comprised of a graviton separator, a sorption cleaner, and filtration with activated carbon for hydrocarbon removal. After purification to 0.3 mg kg^{-1}, the water is discharged or used for irrigation. This process, in place since 1981, is quite slow due to the slow flow rate from the bore holes. A more efficient system is planned which will eventually replace the existing one. Vertical drains have been placed perpendicular to the flow of the groundwater to catch contaminated water. They are designed to lower the groundwater level by 1 meter. These drains will help to minimize the number of pumping and filtration stations. The cost of these drains is high, but they will be required for proper remediation to allow this site to be used in the future as an air base. They will provide lasting protection for future airfield operations.

Nemsová

Contamination in this area was widespread (32,000 m^2), but primarily limited to the surface, or to a depth of about 0.2 meters. However, some sites were contaminated to a depth of up to 2 meters. This site is in the second protection zone of the water source of Nemsová, so strict criteria have been adopted for remediation.

Since the end of 1992, pollution sources have been removed, and surface soils have been excavated for decontamination. This process was undertaken on-site using bioremediation techniques in a concrete-lined basin. Contaminated objects and concrete were cleaned with water vapor to remove petroleum products. Some objects were bioremediated along with the soil. Although a level of 500 mg kg^{-1} was achieved, only pure soils were used as fill material at this site. Complete purification of groundwater has not been achieved at this time.

Lest'

This site was used as a troop training area by the SA. Ten out of 18 sites at this installation were recommended for remediation. Soils at this site were required to have a level of 400 mg kg^{-1} NES, and groundwater was to be purified to category "B" of the Ministry of Environment Recommendations. Technical aspects of this cleanup are similar to that at Sliac-Vlkanová.

Cost of this cleanup is about 237 million Sk, about 25% of the total cleanup budget for all locations. The 1997 goal will most likely not be obtained. Cleanup activities here were negatively affected by the split of the federation and the related demands on the resources of the Slovak Republic. There are also problems caused by new Slovak waste management legislation.

Springer-Verlag
and the Environment

We at Springer-Verlag firmly believe that an international science publisher has a special obligation to the environment, and our corporate policies consistently reflect this conviction.

We also expect our business partners – paper mills, printers, packaging manufacturers, etc. – to commit themselves to using environmentally friendly materials and production processes.

The paper in this book is made from low- or no-chlorine pulp and is acid free, in conformance with international standards for paper permanency.